DR[illegible] THROUGH LIFE

A light-hearted look back on
sixty-five years of motoring

Brian Darwent

Michael Terence
Publishing

First published in paperback by
Michael Terence Publishing in 2020
www.mtp.agency

ISBN 9781800940703

Front Cover image
Karolina Robinson Designs

Contents

1

1939 AUSTIN CAMBRIDGE

Double-declutching, and all that

It's more than sixty years since I first sat behind the wheel of a motor car. In that time, they have changed enormously, as have roads and driving conditions in general. I don't claim to be an expert on cars, and I've never driven one that could be called classic. This isn't that kind of book. I'm simply looking back on an average driving life, and at the vehicles that were for a time under my control. Call it nostalgia.

We had motorbikes first in my family, in the early 1950s. I never got to ride one. They belonged to my father and elder brother. My brother then started courting and a motorbike wouldn't do. So a motorbike-and -sidecar came next, followed by a Reliant three-wheeler. Late one night my brother took a tight bend too fast in the Reliant and turned it over. It had a soft top, and was a write-off. He was lucky to survive. It was then that we bought our first family car, a 1939 Austin Cambridge, at a cost of £100. The year was 1956. I was seventeen and old enough to apply for a provisional driving licence. Mine was the first generation, I would say, able in large numbers to drive from such a young age. My dad was nearly fifty when he started.

It was a peculiar period. We were in the middle of the Suez Crisis. There was an acute shortage of petrol, so the government brought in a temporary rule. Learner drivers were allowed to drive unaccompanied by an experienced driver, though **L** plates still had to be displayed. How this was going to help the petrol situation, I don't know, but the shortage was real enough. I can remember, for the only time in my life, having to syphon fuel from one vehicle to another by sucking on a hosepipe. This could be tricky. The taste of petrol has stayed with me ever since. But my brother soon got married, and my father never became a confident driver, so the car was usually available for my use. Lucky me!

The Austin Cambridge, a conventional, four-door saloon of its time - black, as just about every car was then - had many peculiarities, as they would be seen today. It had mudguards. Lamps were stuck-on bits instead of being part of the bodywork. You gained access to the engine from the sides rather than the front. It had rubber-covered running boards as you stepped in. The interior was quite plush, with leather seats. One useful feature it had was that two of the doors swung backwards, away from you, as you opened them. This made it easier to climb out of than is the case with a modern vehicle. You had merely to swing your legs out of the car then stand up. What could be easier? Unfortunately, the two doors in question were the rear doors, so there was no advantage for the driver. He had the same little struggle that we have today. But the idea was later abandoned, for safety reasons. A door swinging backwards would catch the wind if it was accidentally opened when the car was moving. My brother tells me that he once had a Renault with four doors hinged like that. His son, then a child, almost fell out after accidentally opening one.

The Cambridge's year of manufacture was well before what were first described as winkers. Why they were slow in coming along, I don't know. It is surely an easy matter to cause a light to wink by operating a switch. They were simple, round lamps as well, not the elaborate clusters we see today. The Cambridge had indicator arms. These were normally retracted into the central pillar, but came out, pointing horizontally, when activated by the driver. They lit up, too, though they were much harder to see at night than a winking light. And they had another fault. In cold weather especially, they often remained stuck out. The driver would be unaware of this, however, so when he struggled from his vehicle he would hear, too late, a little cracking noise as his shoulder brushed against the arm, which was made from cheap plastic and easily broken. Gateposts were another hazard.

Using these little arms to signal your turning intentions was a luxury, though. You couldn't use them in a driving test. For that, and to satisfy the highway code, you were supposed to drive with the driver's window wide open. This was so that you could make hand signals. If you were turning right, then obviously you stuck your arm out straight, with fingers extended. Turning left was more difficult. Your arm had to be out, but bent at the elbow. You made a circular motion with your hand, your forearm pointing forwards, more or less. Then there was the signal for slowing down. Stop lights were by no means universal at this time. You lifted and lowered your hand, your palm facing towards the car's door panel. You could also signal your intentions through the windscreen to a policeman on point duty - and we don't see them today - at busy intersections. All this was well-illustrated in the highway code, though hand signals were impossible to see in the dark.

Sometimes, of course, when you weren't anticipating having to make a turn, you could wind the handle and raise the window, to keep out the weather. But the Cambridge's windows were temperamental, or at least the driver's window was. At times it would disengage itself from its mechanism and disappear into the bowels of the door panelling, where it would have to remain till you had a chance to drag it back up. So you got cold and wet, anyway. If you were quick enough, though, you might just manage to catch it in time. What was the car like to drive? Well, it had the normal controls, in their standard positions - pedals, gear lever, handbrake. But driving wasn't so easy, particularly gear changing. In 1939 synchromesh gears - which now make changing gear so smooth and effortless - were rather a new thing. There was no smoothness in the low gears with that car. You had to double-declutch, as it was called, or you would noisily crash the gears. The technique was to depress the clutch pedal twice, revving the engine when the gear lever was briefly in neutral to try to bring the gearwheels that had to mesh together up to the right speed for a smooth change down. It was a somewhat hit-and-miss business, needless to say.

Another peculiarity was that, unlike more modern vehicles, the car didn't have independent rear suspension. In other words, instead of the rear wheels moving up and down separately as you drove over bumps, they moved together. This wasn't much noticed on a straight road, but if you happened to hit a bump on a bend the car would leap sideways by a foot as both wheels left the road surface for a moment.

But talk of driving the car presupposes that you'd managed to get the thing started. Flat batteries were a constant problem. There appears to have been a huge step change in battery technology, but back in the 1950s they could flatten unexpectedly, with nothing to warn you until you turned the ignition key, pressed the starter button, and nothing happened - except for a feeble whirring noise from under the bonnet. Or if it wasn't the battery, it was the choke. They're a thing of the past, as far as drivers of today are concerned. The choke was fitted into the dashboard. When the car was cold you had to pull it out - to enrich the mixture, as they used to say. But this again could be hit-and-miss, with the result that the engine could end up flooded. You then had to sit around waiting for the surplus petrol to evaporate before trying again.

But if you still couldn't start the engine all was not quite lost. In those days cars used to be equipped with a starting handle. This you threaded through a hole at the front, beneath the radiator, where it would engage mysteriously with the engine's or gearbox's innards. You then turned, or cranked, the handle, and sometimes - after it had kicked you back a few times, out of spite, because it preferred to sleep - the engine would splutter grudgingly into life. It was hard work, though you were likely to have more stamina than a battery.

If the starting handle sadly failed, you were left with two possibilities. Here you were lucky if there was a handy hill and you had a companion. One of you would sit in the driving seat, with the clutch pedal depressed and the gear lever in first, while the other pushed the car to the top of the hill. Once momentum was achieved, the driver would release the clutch. Depending on the length of the hill this could be tried several times. And the method was frequently successful - more successful anyway than the starting-handle method, and less laborious, too. The driver would rev the engine, partly in triumph, and partly to discourage it from dying again. If there was no hill, well, you could only try and round up enough people to push the car until sufficient speed was achieved. This wasn't very difficult, surprisingly. Everybody knew what it was like to be in that sort of situation, and willingly put their shoulder against the stubborn vehicle. So one way or another, with luck you would finally be off. It should be borne in mind, though, that it would probably be pitch dark and raining, and you would be miles from home.

As for calling the AA or RAC if you had broken down, not only were there no mobile phones back then, but few people possessed a landline, so there wasn't much point in knocking on doors in the hope of finding a total stranger who might be kind enough to let you use his phone. The motoring organisations in fact had emergency huts in odd places along the highways, and if you were a member you were provided with a key. Inside there would be a phone, I imagine, and possibly also emergency rations. But the chances of breaking down within walking distance of one of these huts seemed remote. The AA also had uniformed men patrolling the roads on a motorbike-and-sidecar. They were friendly people, who if they spotted a member, displaying an AA badge, would salute him. But again it seemed unlikely that such a heroic figure would happen to appear out of the gloom at just the moment when he was badly needed. So all in all, joining the AA or RAC was hardly worth the bother, even if you could afford it.

It was certainly a more chivalrous age. *The News of the World* had a feature, I remember, called 'Knights of the Road'. I'm not sure how they got their information on the gallant drivers who had come to the rescue of some poor stranded soul, though they couldn't have done it by phone hacking. I took my driving test in the old Austin, as did my father and brother, so it was certainly a useful family car. It took me three goes. I don't think driving tests have altered much since then, though they may have been more difficult. The rear window, in particular, was very small, so reversing wasn't much fun. Then there was all that double-declutching, and the hand signals. When the examiner told you to turn at the next corner you had a lot to think about.

And a lot to do.

One unusual thing was that for each of my tests I arrived alone and drove off afterwards alone. Whether the examiners noticed this, I can't say. I was simply observing the temporary rule that permitted it. I finally passed in September 1958, never having had any formal lessons. People rarely did then. Whether the temporary rule was still in operation, I wasn't in truth sure, but I hadn't heard or read any announcement that it had been dropped, the Suez crisis having passed into history. For all I know now, it may simply have been forgotten about. So technically it could still be the law of the land.

2
1948 MORRIS SERIES M
Temper tantrums in the smog

This was a family car again, like the old Austin Cambridge. When I watch old black-and-white films on television, I look out for models I was familiar with at the time. I'm sure I've spotted the Austin, though never this Morris. I fancy it was something of a rarity.

We must have traded in the older car for it, though I can't imagine that we got much for the Austin. It would have been twenty years old. It had faults as well that I haven't mentioned. For instance, the fan used to break frequently. But break isn't the right word. The fans used to come apart - owing to centrifugal force as far as I could ever work out. They were made of some cheap material that was neither metal nor plastic, but something in between. You'd be driving along and suddenly there would be this tremendous bang under the bonnet. There was no need to have a look. Fortunately, it was possible to drive the car slowly without the fan if you gave it a rest every so often. We kept a supply of spare fans at home. The Morris had its faults, too, of course - one in particular. But I'll come back to that… So, more about the road and general driving conditions in the late 1950s We complain about potholes now, but the roads are certainly smoother than they were then. We lived two miles outside Warrington, and close to the main Manchester-Liverpool road, which despite being an important route between the two cities was cobbled on this stretch. You sometimes see cobbles today, preserved in some village because they look quaint. I don't know how horses used to find them, but this main road also had deep grooves where tramlines had once been. It was a terrible surface to ride a bike on, and bumpy in a car. On the drive into town there was one bend in particular where the old Austin used to do its sideways leap. I would keep it as a surprise if I had a passenger with me.

Warrington is built around a crossroads. Besides the Manchester- Liverpool traffic, it was on the main North-South route on the west side of the country. The Mersey could only be crossed otherwise on the old Widnes-Runcorn transporter bridge, to the west, which carried about a dozen vehicles at a time, or via the Barton swing bridge, near Salford, which was a nightmare bottleneck. The town centre is pedestrianised now, of course, but the amount of traffic that had somehow to squeeze through Warrington's narrow streets was huge. The narrowest of them was Sankey Street, where buses had to mount the pavements to pass one another.

And then there were the rush-hours. You might almost say it was the age of the rush-hour. Factories and offices used to disgorge their employees onto the streets at the same time. No attempt was made to stagger the times, and nor had flexitime yet been invented. In fact it was decades away. Some caught special buses, some rode bikes, some motorbikes, but cars were definitely making their presence felt more and more. It would be a long time, however, before anybody seriously questioned any of this.

We had fogs, too, in those days. We don't get much of that now. What we have is little more than mist. Even on quite bad days you can see a hundred yards or so. But in the late 1950s we were still in the age of coal fires. The buildings were black with the smoke and a new word was invented for the kind of fog we had. It was called smog. People were encouraged to wear smog masks when they were out in it. Few ever did, though. It would have been considered sissy. Schools often let their pupils out early, so they could walk home if the buses weren't running. They didn't complain.

Yes, the smog could stop the buses, and of course, driving a car in smog was a nightmare. You couldn't see the white line or the cat's eyes when it got really bad. Or the kerb, for that matter. Windscreen wipers only made it possible to see the swirling stuff itself more clearly. Your headlight beams disappeared into it. To make even slow progress you needed a driver and a passenger. The passenger would wind down his window and stick his head out, reaching across to steer with one hand as he followed the kerb, while the driver just operated the pedals. If he glanced in his mirror he might notice a procession forming behind, for it made life much easier for a driver to latch onto a vehicle in front. There was a risk in that, however, for the driver at the head might lose his way. It was common enough for such a procession to wind up in a dead end, such as a farmyard.

Parking in those days was surprisingly easy, in daylight, despite the congestion. It's true that there were no multi-storeys, or car parks of any kind for more than a few vehicles. But then you could park anywhere. There were no yellow lines, single or double, no parking meters. But after dark was a problem. Whenever I drove into town after dark I headed for a little car park behind the old Empire cinema. It had only a handful of spaces, but I could usually get on it. The problem at night was that you had to leave your lights on when parked on the streets. It was the law. This led to flat batteries, but the police were ruthless in enforcing it. You could buy special little lamps that you plugged into your cigarette-lighter socket and clipped onto the top of your window, but it only gained you an hour or two. Batteries, once flattened, never fully recovered, either. They had something like a bad heart afterwards. The police were also very keen to book you if you parked with your nearside wheels up on the pavement in order to create extra space on the road for other vehicles. We see it all the time now. But I'm forgetting about the Morris. That's because my memories of this car are much less clear. For most of the two or three years we had it I was away in the Merchant Navy. But it was a black saloon once more, with similar features to the Austin - except that this time it was the front doors that swung backwards. I'd forgotten that until my brother turned up a photograph of the model with the handles showing clearly. The car had a funny heating device that you plugged into the cigarette-lighter socket again. Useful, that, though I've never smoked. The device had rubber suckers, which enabled you to fix it to the windscreen. You could make a clear patch with it when you were frozen up. Proper heating systems hadn't yet been introduced. It had been normal enough until fairly recently, in fact, to put a car away for the winter - under tarpaulins if it had to remain outdoors, as if it were a hibernating animal. Cars were once a fair-weather means of transport. We'd built a garage at the back of our house from timber and asbestos sheet, though it was as cold inside as out.

One other useful feature the Morris had were knobs on the inside, attached to the windscreen wipers. When you set the wipers going the knobs turned with them. So how was that useful? Well, if it snowed the wiping area soon shrank because of the build up on either side, so that you were soon trying to see through a crack. But you could twist these knobs by hand to keep a clear space.

I had my first trips into Manchester in that car, and it's amazing now to remember that on a Saturday night I could almost guarantee finding a parking space at the front of the library on St Peter's Square, or failing that in Albert Square, near to the town hall. I also drove into Liverpool in it when I signed on for the Merchant Navy, and I have a memory of that day. I was running late, but I had every confidence that I would find a parking space with ease. This was only 1960, after all. Well, I came upon a big car park not far from the office, the first one I had ever seen, with spaces for at least a hundred cars. And the thing was full. I must have managed to park somewhere, but to be in difficulties of that sort was quite a new experience.

The berth where the ship I was to join tied up was way down the Dock Road. This was cobbled and bumpy. When I drove along it in the Morris later, the car would often exhibit the fault I remember it best for. In places where it got seriously bumpy, the steering wheel would twist violently in my hands. I found it more embarrassing than frightening, especially if I happened to be giving someone a lift. My brother remembers that it was caused by not keeping the steering properly greased, with the aid of a special gun. If you neglected to do this it could become lethal, he says. Well, he's always been the more competent engineer, while I've been more of an artist. What used to happen struck me as more like a temper tantrum. I knew all cars had their idiosyncrasies. I had little affection for this one, so it was simply showing its feelings. That's all.

3
1956 MORRIS MINOR
The leaky car you had to rock

We bought this car, my father and I, when I left the Merchant Navy in 1962. The reason I left was because I had joined as an alternative to National Service in the forces, and that was now at an end. I needed a car for both work and social life. Large numbers of young men were returning to civilian life, and back into jobs. So jobs were scarce. And they were also returning to places where girls might be found, such as dance halls. So there was greatly increased competition in that field, too. Being mobile was essential if you hoped to compete.

You still see a lovingly preserved Morris Minor on the roads now and again, though it had been around for years when I started driving one. It's regarded as something of an icon - our equivalent of the VW Beetle. There are Morris Minor clubs still, I believe. Enthusiasts travel as far as the continent to get together and swap their stories, I've heard. It all sounds very friendly, but it's also a mystery. At least to me.

What different features did it have from the two cars I had so far driven? Well, the doors swung in what we would now regard as the standard way. The lamps were embedded in the bodywork, instead of looking like stuck-on extras. There were no running boards. But it was cold and lacking in comfort, or at least the 1956 model was. If there is one still around you can spot it because the windscreen had a central divider still. At that time it wasn't possible to make curved windscreens, or windows.

Every piece of glass was perfectly flat. And although cars in different colours were starting to appear, this one was the usual black. Of the many cars I've driven and owned, it wasn't a favourite. Rather the opposite. I wouldn't have enjoyed driving to the continent in it. Driving just to work was no great pleasure. It was some weeks, though, before I could do that, so fierce was the competition for jobs. I had served an unorthodox apprenticeship as a draughtsman, and didn't really know if I was properly qualified. The people who interviewed me were inclined to be sceptical, wherever I drove to. The situation looked hopeless. But in the end, I was taken on by a little back street firm in the nearby town of Leigh, for a measly £10 a week, though I told them in the interview that I would gladly work for a spell for nothing, so keen was I to get back into draughting. Eyebrows were raised, but they must have been as desperate as I was.

And so I began driving each morning to Leigh, which could be done partly on country roads. All went well until one day I was held up in mid-country by lights. A broad swathe of land had been dug up. A major project of some kind had evidently begun. I felt indignant, even so. What the hell were they playing at? There was no explanatory sign. No one to ask. A crude wooden track had been laid, so I was able to cross the morass when the lights changed, but it was still a nuisance. I swore at the distant bulldozers and diggers. But then one day a bridge appeared, as if by magic. The whole thing was still a mystery, though. What on earth were they doing? I asked people at work, but they had no explanation. I had seen nothing on television, or in the newspapers. Weeks passed before I found the answer. It seemed they were building something called the M6.

The first stretch of motorway, opened in 1958, would

become part of the M6 eventually. I recall a brief television report. It was referred to simply as the Preston Bypass, with no great fanfare. Motorists weren't waiting impatiently for the day when these amazing new roads would become commonplace. The M1 was opened in the following year, by the Minister of Transport, Ernest Marples. The event was televised, too, but there was still no excited feeling of a huge revolution in motoring round the corner. The M1 seemed remote to anyone living in the north. It was all so very different from the fuss surrounding this high-speed train project. The motorway work proceeded quietly - but very quickly. I first saw this evidence that the M6 was under construction in 1962. By 1965 the new motorway was up and running. But to get back to the Morris Minor, it had its faults, like the previous cars I'd driven. The most obvious one was leaking oil. You hardly see it at all now, but the drive of anyone owning a car in those days would have a large dark oil stain where they parked it. All cars leaked oil a little. The Morris Minor was just one of the worst offenders. I used to park the car in the same spot at work, and the road was black under it. You might think about having the seals changed, but you would be wasting your money. The only thing you could do was keep checking the level and topping it up. Modern cars neither leak oil nor go rusty. So some things in life have definitely improved.

But the Morris Minor's other fault was far more serious. Quite often, and at the most inconvenient times, the starter motor used to jam. As with a flat battery, there was no warning of this. You'd press the starting button, and nothing would happen. Not even that sad, feeble whirring noise you had with a dying battery.

Since it was very unlikely that you belonged to the AA, there were two things you could do. There was a square nut thing on the end of the motor - put there for this very purpose - which you could attempt to turn to release it, if you had the right sized spanner, and if it was still daylight. But of course this invariably happened after dark, and miles from home. So you had to try the other method. It was simple enough in theory. You put the car in first gear and released the handbrake, then you rocked it back and forth - the whole car - which was harder than side to side. It needed two people. If you were lucky, after several minutes of this you would hear a noise - a sort of loud, springy click. If that failed to happen, well, you were simply stuck, a common enough situation with cars in the good old days.

I remember one dark night driving a girl home I'd met at a dance. We were supposed to have coffee when we got there, but I parked the car in a quiet spot on the way, as you did. When it was time to drive on, the Morris Minor of course let me down. I tried to explain to the girl what had gone wrong, but she wasn't listening, and was plainly beginning to lose what little interest she may have had in me. Nevertheless, releasing the jammed motor was a job for two people, so I insisted that we rock the car together, if I was ever to get her home. She was almost ready to walk the rest of the way, but at the last minute, wearing perhaps her best frock, she agreed. And so, in that dark place, we rocked the car back and forth in harmony. There was nothing remotely romantic about it, though I hope the car enjoyed the experience. At any rate it at last produced that longed-for click, so we were able to climb back in and move off. I need hardly add that There was no coffee for me that night.

4

1961 FIAT 500

A case of pointless downsizing

Much later in my driving life, working for a very big company and often out on business trips, I used hire cars a lot. I picked up my phone and quickly made the necessary call. I wasn't concerned about the model it would be. That's because all cars had become more or less the same, at least inside. Similar data was fed into similar computer programmes, with the aim of achieving the most efficient use of materials, producing a similar result, even in different parts of the world. So you hardly knew what make of car you were in. I rarely checked. I would memorise the colour, in case I forgot where I'd parked it, that's all.

But in my early driving days it could take weeks to get used to the feel of a different vehicle. They'd been designed by different individuals in different drawing offices. Little thought had been given to what was on the drawing board elsewhere, and certainly not abroad. So climbing into a different car was a very strange experience. This strangeness was slow to wear off. In the case of the Fiat 500 I don't believe I ever became fully acclimatised.

I'm seeing Fiat 500s on the road again lately, and wondering why it's been resurrected. I feel the same way about the VW Beetle and the Mini. Modern cars are so much better than those old models - far more comfortable and reliable. I'm sure the new versions are much improved, so why pretend to go back to the originals. I suffer from nostalgia, but not for the Fiat 500. Then why did we buy one, my father and I? Economics chiefly, I think. I've known times in my driving life when petrol seemed on the point of being unaffordable. 1964 must have been one of them. And then my working life wasn't going so well. I left the Leigh job under a dark cloud and took up one on the far side of Manchester. This meant either a very long journey by bus, train and again bus, or a slightly quicker one by car. It was a thirty mile drive all the same, and the relatively thirsty Morris Minor - with the extra cost of oil top-ups - wouldn't do. And of course I needed reliability as well. A jamming starter motor certainly wasn't on. And so we bought the little Fiat, though my father wasn't driving much by now, and the important family thing in any case was to keep me in work.

I wasn't working on the far side of Manchester by preference, you see. It was the only draughting job I could find. We imagine now that it was a time of full employment, so you could move easily from one job to another as the whim took you. That wasn't my experience. And this particular new job was definitely a case of jumping from the frying pan into the fire. I strongly disliked it, the work and the people. And I was of course unhappy with that long journey, through rush hours, morning and evening. In fact I soon gave it up, preferring to commute by train and using an old bike for the last part, which I was fortunately able to keep in the left-luggage office.

Perhaps I'm not being quite fair to that little car. I just happened to be stuck with it through a bad period in my life. It did its job. It never let me down, and I rarely seemed to have to fill the petrol tank - though that's inaccurate, because in those days a pump attendant would do the filling, or part-filling. You had the luxury of sitting in your car, paying through the window when he or she had finished. Not everything has changed for the better.

But the Fiat didn't advance my social life. I was now twenty-five, of an age when I should have been moving into more serious courting. The car had a novelty appeal, I suppose, but after a girl had struggled into it once it ceased to be an amusing topic of conversation. More like an embarrassment. It was white as well, with a light-blue interior - cold rather then cosy, besides being so cramped that you could hardly move. As for any thought of transferring to the back seat, well, with its two doors, and a seat apparently intended for small children, that was a joke.

Nevertheless, I did half-seriously court a girl in it for a time. But on top of my other difficulties, she was from Liverpool. A poor choice again, I have to admit. I had that daily commute to and from Manchester, and with a drive to and from Liverpool to follow in the evening, if I was to see the girl. Girls didn't themselves drive very often in those days. Something had to give, and it turned out to be not one thing but two. Yes, I lost both the girl and the job. Even so, I would soon enter a far better period of my life. Somebody up there would smile on me for once.

5

1962 FORD CORTINA MK1

A car to feel proud of at last

This was in a sense my very first car - in fact in two senses. It was the first I bought for myself, without help from my family, and the first I felt proud to be driving. I'd sometimes had a kind of affection for the earlier ones, recognising that I had been lucky to have near-exclusive use of them, but pride was a different matter.

There had been a step-change in car design in the late 1950s, led by Ford and Vauxhall. At the beginning of the decade Fords were still a bit of a joke, with the narrow, cramped, and black Ford Popular as their chief model on the roads. It was as old-fashioned as our pre-war Austin Cambridge. Then came the first Consuls. These were still small cars, but they came in a variety of colours and had an early modern look. It was the much bigger Consul Mk 2 that caught everyone's attention, however. This was a truly modern car. It wouldn't look so out of place on the roads today - in fact you occasionally see one still. At the same time Vauxhalls had brought in distinctive new models, the most stylish of which was the first Cresta. Slightly later, Austin and Morris introduced their new, Farina-designed, Cambridge and Oxford models, but they were never quite a match for their rivals. Other British makes - Humber, Wolseley, Hillman, Singer, Sunbeam - were still to be seen on the roads, but before long the various companies would either merge, their names living on for a time in what became known as badge-engineering, or else they would continue to exist only in old black-and-white movies.

Oddly enough, the Consul Mk 2 was only around for a few years. This was a great puzzle at the time, at least to me. The more luxurious Zephyr and Zodiac models were replaced by the redesigned versions - bigger again - which became the stars of the TV police series *Z-Cars*. The car which took the place of the Consul was this Mk1 Cortina. Mine was to be my first Ford, though by no means my last.

I expect you'll remember this Cortina. Its most distinctive feature was perhaps its rear light cluster. It was circular, and divided into three equal segments. Simple, but somehow memorable. But I haven't explained how I was able to afford such a car, making me the envy of my friends.

Well, the improvement in my fortunes wasn't instant. The fact is that I got the sack from the hated job on the far side of Manchester. I arrived home, dejected, quite late on a Friday afternoon. My situation really did look hopeless. But my mother was having none of it, With time running out, she insisted I go down to the Labour Exchange in town to get myself another job. And amazingly, that is exactly what happened. I was sent for interview to a large firm within walking distance of my home, which till then I had been scared of because I thought the work would be too technical for me. But I found myself among friends, colleagues from my dodgy apprenticeship days. They spoke up for me. I was in - and on a much better salary than in the previous job. Walking to work at no cost, save in shoe leather, I suddenly had money to spare. And before very long I was able to afford a decent car - albeit secondhand still.

Of course the Mk 1 Cortina doesn't seem half so grand,

looking back, but it promised a real step up in my social life. It was maroon in colour, and polished up very nicely. And it proved to be reliable, too. No longer did I need to feel ashamed to invite a girl to climb in. They were apt to be impressed. And so inevitably, as it now seems, I at last met the right girl and the better life started for me - never mind that she too lived on the far side of Manchester, not so far from the awful firm I'd had such a rotten time with only a couple of years before. It was the beginning of my motoring life proper. In 1966, if you had a new car and new girlfriend you wanted to do a bit of travelling. The young these days will already have travelled widely when they get their first car, probably to a number of foreign countries, as well as the length and breadth of their own. So they won't have the same itch. I had been in the Merchant Navy, it's true, and had even visited South America, but I had seen little of England. Though now twenty-seven, my driving up to this point had been local, more or less. The farthest I had so far ventured was the mere fifty miles to Blackpool. What's more, to enjoy travel you needed someone by your side.

Our motorway system has transformed travel, not only by making it much faster, but also by largely eliminating the need for planning. In the old days you'd have to plan your journey, through a series of towns. This wasn't necessarily easy. There were few bypasses. If you belonged to the RAC - which my father did for a while - they would send you free of charge detailed instructions on your best route, avoiding bottlenecks and such. They must have had a whole department dedicated to this service. Today the motorways take you most of your way. You need to do little more than make a note of the appropriate exit. And of course modern sat-navs mean that you needn't bother your head any longer with distinguishing north from south, or east from west.

Not that I have ever been tempted to use one. I'm closer in spirit to the freedom-of-the-road attitude that drivers had in the early days of motoring, before the war, when there were so few cars on the road, than to the age of electronic gadgetry. Who wants to have a voice in their ear telling them what to do? What's happened to the adventure in motoring? If I drive to Scotland I prefer to choose between the high road and the low road myself. But if you've invested in a sat-nav you'll feel that you have to obey its every command.

Back in 1966, the only motorway we had in our part of the world was the M6. But it took you south to Walsall, where you could pick up the A5, which would take you around Birmingham and Coventry to join the M1, and on down to London. Heading north was more appealing, however, for the new motorway meant that the Lake District was now within reach for us for day trips. My new girlfriend - who soon became my fiancee - and I made several visits that year. We also used the old roads to explore the Peak District and Snowdonia, besides other areas of scenic beauty. It wouldn't have been possible in a car with a starter motor that jammed. Or enjoyable in a cramped little Fiat 500. But the Cortina wasn't really so very modern. It lacked many of the blessings we have today. There were no seat belts. You had to wind the windows up and down still. The heating system was pretty basic. Air conditioning hadn't been thought of. The carpets were made from cheap material. And worst of all, there was no radio or CD player. In fact CDs themselves were still decades away.

There was also no starting handle, of course. They were phased out in the late '50s, much to the annoyance of older motorists. It meant that while the Cortina never once let us down, starting any car on cold winter mornings could be very difficult. You might hear an entire street engaged in what I used to call the dawn chorus. The cars would all be going *rrrrrr… rrrrrr… rrrrrr…* often for hours. It's something else that the modern age has miraculously cured. But since I was now in the habit of strolling to work I wasn't affected most days, though we used to have a little paraffin heater that we put in the garage on really freezing nights, to try to raise the temperature by a degree or two. It's a miracle the lot didn't go up in flames.

My present car, although left outdoors in all weathers, invariably starts the first time I turn the key, even when the temperature is down to minus something in the teens. *Brrrmm… brrrmm… brrrmm…* Lovely.

6

1963 FORD ANGLIA

The rot box and the mysterious stowaway

I don't often feel emotional about my cars, but I felt more than proud of the Cortina. I loved it. Like my mother's insistence that I go out and find another job on the day I'd been sacked from the last one, it was a life-changer. Which makes it all the more surprising that I sold it after little more than a couple of years.

Why? False economy again, I'd say. I was clocking up a lot of miles, what with driving the thirty-odd miles to my fiancee's house and back again most evenings, together with our weekend jaunts. But now we were planning to marry, and wished to save the deposit on a house. The Anglia - only marginally more economical than the Cortina - was supposed to help this along. I doubt, though, if we made money out of the trade-in, so it was a pretty stupid move.

The car was grey, too - never my favourite colour. I'm sure you'll remember the model. It had a very distinctive rear window which you couldn't miss. It sloped the wrong way. Inwards. Fords described it as raked. It was supposed to be an important innovation. Anybody in the rear seat had extra headroom, was the claim. It also allowed a bigger boot lid. And most importantly, it stayed dry when it rained. This was long before the time of rear-window wipers. Even the Cortina didn't have those. No one had thought of such a queer rear window until then. It was incorporated in a bigger Ford of those years, called the Classic. The idea didn't catch on, though, and was never seen again after that, a car which few will probably remember today. So much for the Classic living up to its name.

The Anglia was one of the few cars I've owned to appear in a photograph. It's long since lost, but I remember my fiancee carried it around for a while somewhere in her handbag. It wasn't just the car, of course, in the snap. I appeared in it myself, standing alongside it. And then there were sheep in the background, along with the Cat & Fiddle inn, high on the moors above Buxton, in Derbyshire. We must have been on one of our jaunts. I can understand why I was there - and the inn, which is quite famous. Even the sheep. But why I chose to stand with the car, as if I was proud of it, is a mystery. It was black -and-white, the photograph, as most were in 1966. It suited the car's colour. Something else I remember about the Anglia is that it was more prone to rust than my earlier cars. Rust was a major problem back then, and in fact it wouldn't be cured for a long time. Certain cars were referred to as rot boxes. It wasn't unknown for the first spots to appear before a new car had left the showroom, even when it was supposed to have been undersealed. There were all sorts of products that claimed to be able to prevent rusting, but none of them ever did what it said on the tin. Not in my experience.

I wasn't particularly aware that it was a problem with the Anglia, until one day I just happened to notice that the front number plate had adopted a crazy angle. I investigated the matter and found there was a large hole in the bodywork behind it that the number plate had been concealing. It had rusted clean through. I'd never seen anything quite like it.

It was obviously far too late to brush on one of the rust-prevention products. I investigated, and found there were also products for filling holes, if you had been a bit slow in treating developing rust spots. I was realising that could be a never-ending battle. To repair a hole you had to cut pieces of fibre-glass matting to the right size and shape, place them across the hole one by one, and liberally brush on a special resin both sides to fix them in position. It was a messy job, but the resin dried hard and it was just about possible to create a solid new surface that could be sandpapered and painted. There was a limit, of course, to the size of hole that you could repair in this way. The one in the Anglia was only about four inches long by an inch wide. The result was a bit lumpy, but after I had refitted the number plateI, who could tell? The car was as good as new. i used the Anglia for my very first drive down to London. My fiancee had gone off abroad for a week's holiday, long arranged, with an old school friend. It was an opportunity for me to take my mother for a trip. She had never visited London, and this might be her last opportunity to see all the famous sights. And we would of course be able to make use of the new motorways - though the connection still hadn't been made between the M6 and M1, while the M40 hadn't yet been thought of.

Prior to the trip I called in a travel agent's in town. It was probably the first time I had ever done such a thing. They weren't the places we know today. And people then could be extraordinarily unhelpful. I was looking for information on hotels in London. We didn't have a phone at home, and I could think of no other way of acquiring such information. I asked the man behind the counter if he could help. He shook his head, as if I'd asked for something that couldn't conceivably have anything to do with him. I persisted, pointing out that he was supposed to be an expert on travel, and I couldn't see why that shouldn't include travel in the UK. He gave me a funny look, as if I was some nutter who had wandered in by mistake. Then he suggested I write to The Town Hall, London. They'd be sure to help.

So we set off down the motorway without a plan of where we might stay. I'd been to London by train, though, and wasn't unfamiliar with the Underground, so I decided to look for a place within walking distance of a tube station. We stopped a couple of times at the new motorway services - very basic in those days, but a new experiience for my mother. At the end of the journey I found a little guest house in Finchley, the car having given us no trouble at all.

We were only there for two days, but we rode the tube and saw Big Ben, 10 Downing Street, Buckingham Palace, Trafalgar Square, St Paul's and Westminster Abbey, ending up in Hyde Park. We also took in a show. It had all been a great success. Now it was time, early in the afternoon, to begin our long journey home.

There was of course no M25. We had a drive of about ten miles, through heavy traffic, before we reached the M1. The car continued to behave itself - until we had negotiated the A5 stretch and had just reached the M6 - when there was a sudden loud banging noise from somewhere. It sounded for all the world as if somebody had hidden under the car and was now hitting the underside as hard as he could with a hammer.

We couldn't possibly go on. I stopped on the hard shoulder, half expecting a stowaway to crawl out. I felt under the wheel arches, but could find nothing. So we set off again, hoping that we had somehow picked up something from the road, which had now dislodged itself. All seemed fine for a while, but as we gathered speed it started again. Bang, bang, bang. There was something very deliberate, very human, about it. Someone sounded very angry indeed.

So we pulled in at the next services for a better look. I now kept a torch in the boot - carried since the night when I'd needed to release the Morris Minor's starter motor in the dark with the help of a girl. I lay on my back and shone it in every corner. Nothing. I felt around again, but everything seemed entirely secure. Then I happened to touch a front tyre, and realised I hadn't checked those. I wasn't expecting to find a fault because they were fairly new, even if I had bought them from a rather dodgy dealer. I felt all around all the wheels, still puzzled, until I discovered a very long, loose strip in the middle of the front tyre on the driver's side. It had apparently been slapping the underside of the wing, once I'd reached a speed for centrifugal force to loosen it. The dodgy dealer had sold me a re-tread. You don't hear much about them these days. New treads were bonded onto old tyres. They were a common enough thing in the '60s. I could have changed the wheel, but felt we would be safe enough if I took it slowly. Which we were. Nevertheless, when my fiancee and I decided we would drive down to London for our honeymoon, early in 1967, we decided to rent a car. It was a Cortina.

7
1965 FORD CORSAIR
The pirate ship that beeped and died

Nowadays I keep cars for reasonably long periods - three years at least. And I'm much better off these days. But back in the 1960s I seem to have had an itch to trade them in more quickly, regardless of my emotional attachment to them. So it was that, having bought the Anglia in '66, I sold it the next year, not long after we were married. Of course it wasn't a great favourite of mine, but then I hadn't held on to the Cortina for very much longer.

Does anyone remember the Corsair? Why I bought one eludes me. The model was soon snuffed out. There was only the one, while the Cortina went on to Mk 2 and Mk 3, before it was replaced by the Sierra. What's more, the Corsair was of roughly the same size as the Cortina. What was supposed to make it different? What advantage did it have? The cars didn't look alike, it's true, though you wouldn't call the Corsair more stylish. In profile, the front end came to a point. I always thought that it looked rather like a boat. Or an amphibious vehicle, ready to plunge into the waves.

The car was green, which was better than grey or black, but not a special sort of green. Just ordinary green. Nor did it look any better when you'd spent half an hour polishing it.

Was it the name of the car I liked? I can be swayed by names.

But I doubt if I even knew then that a corsair was a pirate, or a pirate ship. I'm sure it never occurred to me at the time. I've said it looked a bit like a boat, though I somehow doubt if the makers chose the name for that reason. It probably came out of an early focus group, with no one caring what the word meant but only how it sounded. Not that it sounded particularly appealing to me. But now I suddenly remember the reason why I bought the car. It was an automatic, you see. A Westinghouse automatic. It had the name on a badge on the back end. I knew Westinghouse was an American company - something to do with aviation? Well, if Westinghouse could make aero engines, or whatever it was that they did, they could surely make a good, reliable automatic gearbox for a car.

Why did I want an automatic? I didn't myself. I acquired it with my wife in mind. She had tried more than once to learn to drive. Now she was talking about it again. I thought if the car was an automatic it would help. You didn't have to double-declutch any more, or stick your arm through an open window to signal your intentions to other drivers, though you still had to change gear.

But she didn't take to the car. Or perhaps she didn't take to me as her instructor. I could be too technical, believing for example that a learner driver should know why it was necessary to change gear at all. She did pass her test five or six years later, but that was after she'd had proper lessons. In the meantime I'd found out that she wouldn't have been allowed to take a test in an automatic, anyway.

And an automatic isn't necessarily easier to drive in any case - not if you've been using a manual gearbox for years. The Corsair had a conventional accelerator pedal, but the brake pedal - the only other one needed - was a horizontal bar, nearly a foot in length. It wasn't difficult to work out why the designers had provided it. Their idea, I'm sure, was to give you the choice of using either your right foot or your left to brake.

How very thoughtful of them! But there was an obvious danger. You could tell from the note of the engine that the gearbox was about to move up a gear, or down, and it was almost impossible to stop your left foot stamping on the clutch - which of course didn't exist. Instead you stamped on the long bar - the actual brake - and the car came to a juddering, completely unexpected halt. It happened to me three or four times. Horns blared. Fists were shaken. Other drivers thought you'd gone completely mad, and shouldn't be allowed on the road. Luckily, I never caused an accident. I don't know how. Who is ever ready for the vehicle in front to screech to a halt in front of them, for no reason whatsoever. Until I'd got used to the car, I would tie my left leg to the underframe of the passenger seat, keeping it well out of the way of that long pedal. And even then I could feel it tugging whenever an automatic gear change was imminent. I'd bought the Corsair from the main Ford dealer in town, for once. I usually got my cars out of newspaper advertisements. Perhaps an automatic was hard to locate. It was the first time I had dealt with them. How different main dealers could be back then! Let us say they could be unhelpful. The idea that you were the customer, and they were there to serve you, hadn't yet caught on. They didn't give a damn how long they kept you waiting at a counter, and appeared to shrug their shoulders if you were dissatisfied with their service. There were plenty of other fools out there keen to buy their wonderful motor cars, they seemed to be saying.

But the Corsair was reliable enough to start with. We had our first trip up to Scotland in it, I recall, stopping first in Edinburgh, then exploring Perthshire, before driving down the Ayrshire coast. On the way home we had a night in the Lake District. A great holiday!

Soon after that, though, the car developed a problem. Every time I went over a bump the horn sounded. Very strange. And very embarrassing, too. I was receiving some very funny looks on the road. People seemed to think I was a desperately impatient driver. I tried to gesture from my driving seat that it wasn't my fault, though with no apparent success. So I did the only thing possible in the short term. I found the cable to the horn under the bonnet and cut through it with a pair of scissors. Then it was time for a visit to the main dealer.

Of course, immediately after work, there were a dozen people awaiting attention - and being dealt with very slowly. Each customer seemed to need a ten minute discussion. There was no indication of urgency, though they would be closing at six. But at five-to-six I at last got to the head of the queue.

I explained, while the man examined his watch, what the problem was. He didn't seem very interested, but wrote my details on a pad, and appeared to half take it in when I stressed that I had deliberately cut the cut the cable, so just joining it up again wouldn't do.

Two days later I picked the car up, handed over an exorbitant amount of cash, and set off for home. All went well until I went over the first bump. And sure enough the horn sounded - just as before. I lifted the bonnet, and not really to my surprise I found they'd joined up the cable again. Nothing more - though in fairness they'd made a very neat job of it.

I drove straight back and made a great fuss. Nobody would listen to you unless you shouted. I had to wait a couple more days, but this time the problem had been fixed - and at no extra cost. But that wasn't the end of my problems with the Corsair. Nor of my arguments with the Ford main dealer. After about eighteen months, the car one day simply refused to move. The engine was running, I'd put the gear lever into the correct position, but the car simply stayed where it was. There was nothing I could see under the bonnet that might explain what was wrong. The car seemed to have died - while still breathing.

I had to visit the Ford dealer by bus. When I explained to the man this time that the car had a Westinghouse automatic gearbox, he didn't seem very surprised. They often gave trouble, he told me. I asked what could be done, thinking that the car would simply have to come in for repair. He said that nothing could be done. They no longer had dealings with Westinghouse. That seemed to be the end of the matter as far as he was concerned. I could see he was about to turn to the next customer. It was time for a really huge fuss.

The manager was brought in. He too said that nothing could be done. I told him I ought to have been advised when I bought the car that any trouble with the transmission would not be reparable. It took some time to get him to see the fairness of this, but after I had shouted loud enough for everybody in the building to hear, and threatened to contact the local newspaper with the story, we struck a deal. He told me to come in next day, when I would be able to choose from their extensive stock of used cars a model of comparable value, at no cost. The company believed in complete customer satisfaction, he said.

8
1965 FORD ZEPHYR
The Z-car I didn't choose

Would any man today think of choosing a new car without taking his wife along - or at least discussing the possibilities with her beforehand. Yet that is precisely what I did. It's true that she was some time away still from driving herself, but really…

It was the manager himself who had agreed to meet me - and he was there waiting. I had yet to tell him that I needed to have the Corsair removed from my drive and somehow disposed of, but I felt it was best to leave that for the present. Perhaps the reason why I hadn't involved my wife was because I had no idea what would be on offer. I was secretly hoping I might be able to get my hands on another Cortina. I could see one or two in the line-up, but the man told me they were the wrong year. He didn't appear to be very positive about any of the cars we could see, and I began to worry that he was planning to fob me off with something inferior, hidden around a corner.

Then we came to a big maroon Zephyr. He strolled around it and seemed unwilling to move on. I wandered farther along the line, but he stayed put. It dawned on me that he'd decided in advance which car I was going to take away.

I hadn't dreamt of choosing a Zephyr. I thought of it as a police car, because of *Z-Cars* on television. It looked huge. Would it fit in our garage at home? I looked in through the window and saw that it had a steering-column gear-change lever. I had never used one of those. It put me off. I simply didn't need such a big vehicle, anyway. But the man's mind was obviously made up, if mine wasn't. I could see that if I carried on resisting, things might well get more difficult. What was my legal position exactly? Did I even have one? Then I noticed that the car had seatbelts. They weren't yet compulsory, but a campaign to encourage people to fit them was underway. It would be useful if a car already had them, since it might soon become the law. My wife was also pregnant now, and I was becoming more safety-conscious.

So the seatbelts swung it. I nodded to the man. He seemed very relieved as we went inside to sort out the paperwork. It was the moment to raise the question of the Corsair, still standing immobile on the drive of our new bungalow. He didn't put up much of a fight. There have been a number of campaigns in my lifetime to persuade people either to change their habits or to accept something new. It didn't work with smog masks in the 1950s, but it has done with smoking, and is making progress with drink-driving. Offices and pubs used to be full of smoke, while you thought nothing of climbing into your car after drinking three or four pints of beer. And among the more successful of these campaigns has been to persuade people to wear seabelts. The campaign was fronted by none other than Jimmy Savile. *Clunk, click, every trip!* Remember that?

Given some support from the motoring organisations - who also opposed speed limits - many people were against them. It was partly because they didn't like being told what to do, but there were stories too of drivers being unable to release themselves after a crash and as a result perishing. And again, some people couldn't understand why they were needed at all. I found it hard to explain to my mother that if the car stopped suddenly she would be thrown forward at the speed at which it had been moving. She was convinced that she would stop with the car. And after the arrival of our first daughter, in 1970, my wife, who had a university education, believed that if she had her seated on her lap in the front seat the baby would be safer inside the seatbelt than outside. Nor were they alone in being confused. It took imagination to work it out. But gradually people got the message.

It also took time, however, for seatbelts in the back of a car to be accepted as equally necessary. The Zephyr had them only in the front. This remained the case with many cars for quite a while. I suppose the idea people had was that the backs of the front seats, together with the backs of people in them, would form a sort of safety cushion in a crash. All the talk was of drivers being crushed against the steering wheel, or even being thrown through the windscreen, if they weren't wearing a seatbelt in the front. Then came tales of passengers being projected through the windscreen from the rear seat, and that made everybody think a bit harder. But it all took time, as I say.

I had managed to drive many miles without giving any thought at all to my safety. That suddenly felt extraordinarily reckless. I quickly got in the habit of fastening my seatbelt in the Zephyr, despite the first ones being less convenient than now. You had to adjust the buckle to suit your size, like a belt for your trousers, and then if the belt was tight - as you felt it needed to be - your movement was very restricted. The seatbelts we know today, which lock only when sharply tugged, as in a crash, are a great invention. I do wonder, though, how many drivers actually realise how they operate.

What to do about a baby's carrycot, or later a toddler's seat, also focused my mind on safety. I remember that I had to create anchor points for safety belts for these by drilling holes through the floor of the car and tightening bolts underneath. My brother tells me that I used to drive fast when younger, particularly in the Morris Minor. This strikes me as bizarre, although I always seemed to be in a desperate hurry until I got my working life sorted out. But with the Zephyr I believe I really slowed down. I remembered seeing the big American cars from a base near Warrington gliding slowly around. It's possible they were looking for girls to pick up, but it struck me as a kind of luxurious way to drive, if you had a limousine of sorts.

I did have another period of panic in my working life, though. My wife was a pharmacist, able to earn good money, but she took time out from the job when she became a mother. I wasn't earning enough, so I took the reckless step of going contracting. This amounted to temping for draughtsmen. It paid well, but often you weren't called in by a firm to work on a drawing board, but for some stupid task they couldn't find anybody else to do. I once had to follow a plumber on his rounds, then make a schedule of his activities, because he was secretive about them. But then one day I was sent by chance to a big company with a familiar name, which saved my life all over again.

Meanwhile the company I had left behind to go contracting, which had seemed perfectly safe - one of four or five factories on a big industrial site, making everything from aircraft undercarriages to forklift trucks - soon closed down, along with the rest. A big housing estate has stood in their place for many years now.

But I don't have to keep the name of the important company I joined to myself. It was ICI, then in its heyday. I would prosper there for twenty-four years. Had I approached them by the conventional route they wouldn't have entertained me, but as a contractor I could enter by the back door and prove my worth.

But to get back to the Zephyr, while it seemed a big car, it wasn't as roomy as you might think. It was wide, certainly, and it had a bench front seat, which was why it had the steering-column gear-change. The seat could accommodate three people, and was also handy when you had light shopping. And yet, front to back, the passenger compartment was actually quite small. Space for legs was even a little cramped in the rear seat. In those days designers were generous in creating space in the engine compartment and boot, all at the expense of passenger comfort. Transverse engines, which had been introduced with the Mini and greatly reduced the depth of engine space needed, had yet to be universally adopted, while boot space was probably maximised with the luggage needed for domestic holidays in mind. Package tours to the *Costa del Sol* had barely begun.

The Zephyr's boot was certainly roomy, but at the same time not really spacious enough for carrying the things you needed with a new baby, such as a cot, even a dismantled one. So I was soon looking for a more suitable vehicle.

I was driving to work again, and remember that the exhaust pipe fell off one Christmas Eve, on my way home in heavy rain. But bearing in mind how it had been acquired, the Zephyr had given surprisingly good service. Why the manager at the Ford dealers had been so keen to off load it remains a mystery.

9

1969 HILLMAN HUSKY

The night of the salvaged radiator

The circumstances in which I came to possess each new car - and I do mean secondhand car so far - have not always stayed as clearly in my mind as the late afternoon when I chose the Zephyr. But I also have a clear memory of the day I drove this Hillman Husky home, although for very different reasons.

It was a Saturday. My wife and I had been to Liverpool for some reason the week before and had noticed a half dozen of these Huskies lined up outside a dealer's premises, all blue. The Husky was an estate car version, as cars with a tailgate were then called, of the Hillman Imp. In fact it looked more like a small van into which side windows had been fitted. So we stopped and had a look. The engine was at the rear, but if you lifted the tailgate you could push down a part of the rear seat to make a large flat area in combination with the engine cover, which we judged would accommodate all the stuff we needed to carry around. At the same time a small part of the rear seat remained as a seat, big enough to attach a child's seat to, and it would soon be time for one of those. The Husky seemed to be ideal for our needs. So we obtained a trade-in price for the Zephyr and I arranged to collect the vehicle we'd chosen the following Saturday. My father came with me on the journey, just for an outing. It all went smoothly, but little did we know that there had been a drama going on back at my parents' house.

We stopped for a drink on the way home and were late

getting back. There was a police car standing outside the house. We thought something must have happened to my mother, who wasn't in the best of health. But she came to the door. When she'd seen the police car draw up she had almost collapsed, thinking we must have been in a serious accident. The lady next door, meanwhile, had been looking through her window, thinking the same. It was her husband who had died, however. He'd gone to collect his car after a service, and had a heart attack on the street. The police had called on my mother to ask about her neighbour's state of health. They were now next door. Not a day you'd easily forget. The Husky did its job very well. Loading stuff in the back took no time at all. But it wasn't the most reliable car I ever owned. And it was a different age for dealing with car problems, as also for the general maintenance of a vehicle.

It's only quite recently, having had two new cars in succession, that I have given up various checks you used to make as a matter of routine. More than that - checking the oil, water, and tyre pressures was a matter of conscience. They had to be carefully watched. It was part of a man's sense of responsibility. And each time you lifted the bonnet you had a quick look at everything, giving this a tug and that a shake to be sure all was in good working order.

I never seem to see men on their drive, with the bonnet of their car up, today. Or their feet sticking out from underneath it. Even ten- year-old cars look new, and their reliability probably matches this. So the younger generation of drivers appears to have given up on that checking routine. And it's the same when it comes to servicing and repair work. They take their car into the garage, or if once in a blue moon it breaks down they call the AA or RAC. Their masculinity is no longer on the line. I'm copying them now.

But forty-odd years ago you did your best to fix things yourself. I used to belong to a car club in my early ICI days, and one car in the group at least would be off the road each week, meaning that the driver wasn't doing his fair share. No one ever questioned this, however. It was bad form to do so - an unwritten rule that a man must be allowed time to repair his car himself, no matter how long it might take.

I hadn't been in a car club, though, when the exhaust pipe fell off the Zephyr on Christmas Eve, but on my own, in the rain. It hadn't fallen off altogether, of course, but, lying in a wet gutter, I managed to yank most of it away from the piece attached to the engine. Exhaust pipes used to rust and begin to fall to pieces, which at least made the job of removing them easier. I threw the pieces into the boot and drove on, making a tremendous din. That's something else you rarely hear now. We were due to set off for my in-laws when I got home, and we went ahead with the plan. Funnily enough the police, who were otherwise vigilant, didn't seem to bother about noisy exhausts.

We forgot about the car on Christmas Day, then, on Boxing Day, I managed to find an exhaust-pipe place that by chance was open. I bought the new pieces I needed and fitted them, lying on my back under the car. Some Boxing Day! Not only was it an awkward and unpleasant job, it was also at the limit of my technical expertise. My brother could strip an engine, or fit new clutch plates. While I could always draw better than he could, I wasn't very practical. I believe it was after this exhaust experience that I joined the RAC. My card says I've been a member since 1971. * But what about the Husky and its problems?

As usual all went well for some time, then a strange thing began to happen. Or maybe not so strange, in as much as it resembled the old trouble with the old Morris - although coming from the back end, where the engine was. But the whole car would begin to shake and shudder. Only now and then, though. Had it been constant, I would have had to have it taken in to a garage for investigation and repair without further ado. But no, the trouble was occasional, and this time appeared to have nothing to do with the road surface. The car was having a fit of some sort, or so it seemed to my non-engineer's view of things.

I spent weeks trying to get to the bottom of the problem, either crawling about underneath, or looking down on the engine with the cover plate removed. I made no progress because it didn't happen when the car was stationary. It had to be moving. So I drove around with the engine exposed, and even tried seeing what was going on through the mirror. Then I got a friend to sit in the back and observe the shaking from close quarters. He reported that the whole engine suddenly began to shudder. As to what was causing it to happen in such a random way, he really couldn't tell me.

So I finally had to do the thing men in those days hated doing. I took it to a garage. I stayed around while the man had a look. He lifted it up on the ramp and shone a torch. Then he tested each of the nuts attaching the couplings to the rear wheel. One or two were loose. He found a spanner and tightened them. It only took a few minutes. I had already told him that the problem had been driiving me mad. He was sympathetic and said he wouldn't charge me. The shaking and shuddering disappeared. It challenged my belief that garages were run by rogues, out to rip you off.

The other trouble I had with the Husky concerned its

radiator. Or radiators. I had at least three. They kept leaking and spurting steam. It happened once when we were at my in-laws' house for the weekend. We were supposed to be visiting some friends of my wife for dinner, and phoned to tell them what had happened. The husband said he would drive over to collect us. Regarding our radiator problem, he thought he could help. He was friendly with a local garage owner who he knew had a Hillman Imp on his premises that had been in a crash and was a write-off, though the radiator would certainly be undamaged, and would suit our vehicle. The garage man might let us have it.

When we reached his house he immediately phoned the garage, which was also a petrol station, and open. I kept on trying to tell him to leave it, we didn't want to spoil the evening, but there was no stopping him. The phone call seemed less than straightforward, but the result was that we could have the radiator but we would have to remove it ourselves. Our friend told his wife, who was in the kitchen preparing the meal, that he and I would go to the garage right away and bring the radiator back. We could take it to my in-laws' house and we'd fit it the next morning. Sunday.

So we set off. It was a cold January night, fully dark. When we reached the garage he told me to wait in his car. He was gone some time, returning eventually empty-handed. I had expected him to bring out tools and a torch, and thought the mission had failed. I was all for returning for dinner. But he was a man of iron will. He led me round to a dark place, where I could see the shadowy forms of wrecked cars. Somehow he spotted the Imp. Luckily the tailgate wasn't locked and lifted easily. But you couldn't see a thing inside the car. I said to forget it, for the umpteenth time. I was shivering. I knew the radiator was only attached with small screws, and asked if he had a screwdriver.

He held one up. It looked tiny. Then, while I stood back, he set to work. I could hear groans of frustration. Half an hour passed. I thought I might catch pneumonia. Dinner would be spoiled, We would be in trouble with our wives. And I couldn't imagine how he would ever remove a radiator, in the dark, with only a small screwdriver and no torch. But I was wrong. I heard a sound of metal dragging against metal - and suddenly out it came. He lifted it over his head in triumph. I had to congratulate him, and apologise for my lack of faith. He put the thing in the boot of his car, and we returned for dinner, which was just ready.

He drove us home later, and next morning he returned to help me fit it. I expected to have to do no more than stand by while this superman did the work unaided. He lay down on the road surface to inspect something. Then he got slowly to his feet, holding his head. He explained that he had a condition that wouldn't allow him to bend low or lie on the ground for very long because he became dizzy and nauseous. His own in-laws lived a few doors away. He disappeared into their house without another word.

I was left to finish the task alone. The salvaged radiator was a great success, too. I had no further trouble with it. Before leaving the Husky, I mustn't forget a bizarre incident involving it. I've only ever been in one serious accident, and that was back in the 1950s. Five of us had sqeezed into an old banger a friend of mine was thinking of buying. I was in the back seat, hard up against the door on the driver's side. We were taking it for a test drive, in the dark, and we hit an articulated lorry, with headlights full on, coming towards us. Our driver couldn't see that the back end of this vehicle had yet to pass a parked car and was on our side of the road. We ploughed right into it. The door I was pressed against was torn off, and the rest of the car was a wreck, and yet we were all unhurt. Since then I must have driven to the moon and back, but all I've had otherwise are a handful of minor collisions - except, that is, for this weird incident in the Husky.

I was travelling home from work, along a straight, narrow street in a built-up area. I wasn't speeding, indeed I was probably doing no more than twenty. It was growing dark, but up ahead I could see a young boy standing on a corner, outside a shop. I felt sure he would look my way before he crossed, but he didn't. Instead he ran in front of me, leaving me no time to brake. I jammed on, but couldn't stop the car bumping him. It wasn't much, though, and I expected him to get his feet and continue on his way. But he lay in the road.

I leapt out and bent down to see what harm I had done. There was no sign of injury, no blood, but his eyes were shut. It was then that the really bizarre thing happened. Parked only a short distance away for some strange reason was an ambulance. There were men in it and they had seen everything. They arrived at the boy's side only seconds after me. I was ignored. They brought a stretcher, lifted him carefully onto it, deposited him in the ambulance, then drove off. They didn't even ask my name.

The shop, a newsagent's, was still open. I felt I ought to contact the police, and asked to use their phone. After half an hour a police car arrived. I explained what had happened, but they didn't seem to believe me. The man in the shop had seen nothing. I almost had to force them to take my particulars, and asked if they'd follow up what had happened and let me know. But they never did.

So I phoned the hospital and asked if a boy had been brought in after an accident on the evening in question. I was concerned by now that the authorities would have me down as a hit-and-run driver. They took some persuading, but at last gave me a number for the boy's parents. I phoned them, giving my side of the story, a little worried that the father might turn angry. But they seemed oddly uninterested, too. They told me their son hadn't really been hurt, and that was that.

10

1973 VAUXHALL VIVA

Keystone Kops on the M62

The volume of traffic now crosssing the Mersey and the Manchester Ship Canal via the two bridges which form the Thelwall viaduct a few miles to the east of Warrington is an indication of how drastically conditions have changed since I started driving in 1956. Before the M6, no crossing point existed there. Environmentalists tell us that the more roads we build the more they will fill up, so why build them? They are surely right. And yet how glad we are when car production in this country is up, reassuring us that the economy is doing well! There's no answer to this. It's the same with the economy as a whole. We bemoan the increasingly senseless consumerism which has us in its grip, yet if the cameras show us closed shops on high streets and empty shopping malls we worry. Who wants more recession and unemployment?

This relentless build-up of traffic has also occurred on the Widnes- Runcorn suspension bridge, the biggest of its kind in Europe when it was opened in 1960. The old transporter bridge, trundling slowly across the river and canal, couldn't have moved more than a hundred cars a day, so I expect the designers thought they had provided enough capacity for a century at least. So confident were they, apparently, that the steelwork at the ends of the arch encroached into the carriageway. This restricted the width of the road over the full span. But of course after little more than a decade the traffic volume almost engulfed the bridge, causing frequent rush-hour jams. And so work began on modifying the basic steel structure to provide extra lanes. Imagine that! It seemed to go on for years. And among the frustrated drivers, first in the Hillman Husky, then in a Vauxhall Viva, was me, travelling from our home north of Warrington to the big ICI plant at Runcorn, south of the river. We didn't keep the Husky for very long, even though it gave us little further trouble after the shuddering and radiator problems had been cured. My brother was working at the Vauxhall car plant at Ellesmere Port in those days, and he told us they had a scheme through which he could order a new car for us, at a very attractive discount price. The idea was tempting. We had never really thought of a new car till now. They'd always seemed out of our reach. But I liked the idea of a car which would presumably be trouble-free for a couple of years at least - including rusting problems. Meanwhile there was by now less paraphernalia to carry about, so we could go back to a normal boot. We decided to take him up on the offer.

It was while I was driving the Viva that I was transferred to ICI's plant at Northwich, to work on development projects. This made the journey to work even more arduous, so we decided to move south of the river into Cheshire. Both moves turned out to be very positive. I had more interesting work to do, which didn't chain me to a drawing board, and I found myself travelling to work through the countryside. The only serious obstacle in the mornings was the occasional herd of cows.

ICI had its own large workshops, but these were unsuitable for the kind of work I was now engaged in. More specialist expertise was required. The smaller firms which helped us turned gradually into our development partners. This was encouraged. Several of them were in the Midlands. Meetings were often needed to resolve problems. So I found myself on the road pretty frequently. And because I was from ICI, while there was no actual red carpet, I was well-received. I didn't have to cope with shop-floor arguments with in-house fitters and such. I was listened to with attention and respect. And after a meeting I would stay for lunch - taken out to a decent local pub or restaurant rather than having to sit in a works canteen, or even a staff dining room.

I must say that I took to the life. After my earlier work experiences it seemed too good to be true. I was allowed to use my own car as well for the present, with generous expenses. Rules about using hire cars only came in later. Mobile phones lay well in the future. Once I had driven off site I felt free. I didn't yet have a radio in the car, but when that day came I would listen to an interesting programme or play. Returning to site later in the afternoon I would half expect a senior manager to be standing on the office steps, ready to demand an explanation for the time I'd spent away from my drawing board, but nothing like that ever happened. I was trusted. Nor was this trust really misplaced. The fact was that I was very conscientious. Given the freedom of the road I did nothing worse than call somewhere for a haircut once or twice. Talk about corruption! But of course the miles were piling up, and as might be expected the Viva began to give me a few problems. I had a lot of trouble with the steering, I remember. I don't mean that the steering didn't suit me, but that I began to lose confidence in it. It felt loose. When I paid it close attention, I found there was play developing. I could turn the wheel a little each way before it engaged. Steering mechanisms were way beyond my expertise. I could hardly take the car back to Vauxhalls, so I had to find a garage - preferably one within walking distance of my place of work. And it happened that there was such a one, run by a man of unknown nationality.

He had evidently been in this country for a long time, however, for whatever the problem was that you brought him he always began with a ramble about how he had been in the motor trade for twenty- five bloody years. Sometimes he repeated this. It was difficult to get into a detailed discussion of what the problem was and how it might be cured. I was always pressed for time, and since there was nowhere else close by I had to try and believe in him.

It would be nice to be able report that in spite of the difficulties of communication he fixed the trouble. But he didn't. After several visits, the steering remained loose. The car was becoming dangerous to drive. And when I tried to complain I would get the usual stuff about how he had been in the motor trade for twenty-five bloody years. The only thing to do was to hand over the money. In the end I contacted my brother. If anybody understood the steering mechanism of a Vauxhall Viva, he ought to. And he must have given me good practical advice because I managed to rectify the problem myself, at home. My business trips weren't confined to the Midlands. I used to visit a firm over in Yorkshire pretty often, via the new trans-Pennine M62 motorway, and it was on one of these trips that I had easily my most memorable experience in the Viva.

I'd had the usual meeting and lunch, and set off home, too late to report back to base. Not that anybody would be waiting for me. I had just pulled onto the motorway, and was starting to gather speed, when an extraordinary thing happened. The accelerator pedal return spring failed. The engine was suddenly racing at top speed, while the pedal remained lifeless in the hard-down position. What should I do? If I pulled onto the hard shoulder, what then? All I had to control the car were the clutch and brake pedals, but I decided to drive on. The traffic was relatively light, so I was able to weave between lanes, occasionally depressing the clutch or touching the brake. It was like an old *Keystone Kops* movie!

Rather than travel all the way back to Cheshire, I decided to call at my in-laws. I also knew where there was a Vauxhall main dealer. But I couldn't drive like that on the ordinary roads. As soon as I'd left the M62 I stopped and switched off the alarmingly racing engine. I had an idea. I opened the briefcase I carried, hoping there might be a rubber band in it. And miraculously there was - a thick one. I lifted the bonnet. I knew enough to find the carburettor, but not much more. Even so, the broken spring was easy to spot. I removed it, threw it away, and fitted the rubber band in its place.

Back inside the car the accelerator pedal had now risen, as if from the dead. I started the engine and revved it a few times. So far so good. I drove very gingerly to my in-laws' house, explained what had happened, phoned my wife, then set off for the main dealer.

Luckily they hadn't yet closed, but we were in the age of unhelpful people still. The man asked me for the broken spring. I said I had stupidly thrown it away. So he asked what kind of carburettor the car had. I said I had no idea - how many kinds were there? There were two, it seemed. I told him I wouldn't know how to tell one from another. He said in that case he couldn't help, and turned away. I could see he was about to shut up shop but I called him back. I suggested he might bring me both springs and I'd see which one fitted. Very grudgingly, he went to look for them. I was fully expecting bad news, but he brought them both, tossing them down on the counter. I went out to the car and established which was the one that fitted. It took only seconds. I went back inside grinning, all ready to thank him for saving my bacon and hoping he might be glad to have been of service. But as I paid, his expression was as surly as ever. I expect he'd had to work for an extra minute or two.

11
1975 FORD CORTINA MK 3
200 miles home without a brake

The Vauxhall Viva was light blue. This Cortina was the second white car I'd owned. I haven't had a black car since the Morris Minor. You still see a few black cars around. I expect they polish up well. But for those of us who grew up when every car on the road was black there's a prejudice against them.

They say, as you get older, you remember times past very clearly, but forget what happened yesterday. It's rather like that with cars, I'm finding. On the other hand, when I bought this secondhand car in 1977 I was only thirty-eight, so I should remember it pretty clearly. That year is after all a long time ago now. The Queen's Silver Jubilee, was it?

Why did we let the Viva go? It was only four years old. I know our second daughter had arrived, early in 1975. So we had plenty to carry when we went anywhere. What had happened to the need for space for children's paraphernalia that had persuaded us to buy the Husky? I don't know. There was one convenient thing, however. When it came to fitting toddler seats in one of these cars - I believe it was the Cortina - I found that it wasn't necessary to drill holes through the floor. There were plates under the carpets to which the belt anchors could be attached. Indeed, it looked as if they had been put there for that very purpose. The car manufacturers had actually collaborated with the seat makers - Brittax, I recall - to make life easier for car owners with small children. Very little of that sort of thing went on in the old days. To discover that something unexpectedly fitted was against nature. Almost a religious experience.

The Viva was a two-door car. This had seemed fine when we got it - safer with a child in the back. But travelling along the Cheshire lanes to work I was in a car club again. Five of us, picked up and dropped off in different places. There was a lot of stepping out of the vehicle to tilt a front seat so as to allow access to the rear. And of course I spent much time removing and later refitting a child's seat in the back, on my driving days. The Cortina conveniently had four doors. Ever since my infatuation with the Mk1 back in the 1960s I had also had a thing about them, if not about all Ford cars.

What was distinctive about this latest Cortina? It seemed to be low-slung, almost sporty, compared with the earlier model. I was used to climbing into a car, but with this one you sort of lowered yourself into it. Your legs then seemed closer to the horizontal than I'd ever known before. In fact you felt as if your backside was just a little too close to the road surface. And this low-level driving position gave the illusion of extra speed. Fifty seemed like seventy. Of course as someone who'd started out in a pre-war Austin Cambridge, I didn't much mind this - despite having made efforts to curb my speed since the days of the Zephyr. The trouble was, it had the same effect with front-seat passengers. My wife had recently passed her driving test. Before that she'd paid no attention to the road ahead. That was my concern. Now, when I was driving, she couldn't take her eyes off it, as it rushed under the bonnet. She had become a very nervous passenger, in short. There was only one solution. I would have to drive more slowly for real. Our car-club journeys to work tended to be slow in any case. This was because there would be arguments going on amongst the five of us. Sometimes these could become very heated. The other members were all senior to me, but that didn't discourage me. I felt that too much was at stake.

It wasn't really anything to do with work that we were debating so furiously, it was politics. The nation was at a political crossroads as never before. There had been little disruption at ICI, but the country had been passing through a Winter of Discontent, as it has come to be remembered. A Labour government had failed to get a grip on an epidemic of pay demands and strikes. But now a Prime Minister arrived who was determined to confront the unions head on. It was Mrs Thatcher, of course, elected in 1979.

I was a Labour man through and through, and not yet ready to concede that she might have a point. I was also under the influence of the most famous, and readable, economist of the day, JK Galbraith. A believer in incomes policy, Galbraith deplored what Thatcher was doing, and was an equally strong opponent of the policies of the new Reagan administration in America. In so far as I've ever understood it, their idea was to choke off the supply of money at source, so that the pay increases being demanded by the unions simply couldn't be met. The immediate effect was a dramatic increase in unemployment. We seemed to be heading back to the 1930s, and the Great Depression, on which Galbraith had written the standard work. I couldn't see how that could be allowed by any government. It looked suicidal. But Mrs Thatcher still persisted. So that was what we were arguing about as the glorious Cheshire countryside swept by the windows of our car. I was in a minority of one.

And I fear the senior people were right. But the irony was

that the new business order that was soon to be created didn't suit ICI. It threatened the jobs of my travelling companions more than it did my own. The company was a dinosaur, ready to be broken up and sold off to the highest bidder. And that is more or less what came to pass, not so many years later. It used to be described as the bell-wether of the British economy. Now few people remember it. Dominance has passed from manufacturing to the City of London, which ihas proved to be as irresponsible in its way as the unions once were. The Cortina, meanwhile, hadn't been giving me much trouble, though that could not of course last. In the year of Mrs Thatcher's election we had out first holiday abroad, driving down to Luton airport to fly to Italy. All went smoothly until the return journey, then, as we approached the M1, the brakes started making a horrible noise. The pads had evidently worn through. Our youngest was only four. I didn't wish to take any risks, but the least risky thing was to continue by motorway, it seemed to me. I was in a minority of one again, but I argued that it was on motorways that you were least likely to need to brake. If I left a long gap in front I ought to be able to slow down using the gears, should the need arise. Of course if there was an emergency I would have to jam on, and we would be faced with the expense of new discs as well as pads. How often did that happen, though? I was still in a minority of one, but I drove slowly onto the motorway, leaving a gap of about a hundred yards to the next vehicle. I managed to sneak along the stretches of normal roads in similar fashion, too, and got us all home safely without once touching the brake pedal.

The other thing I remember regarding this Cortina is that for the only time in my driving life the windscreen shattered without warning. Or rather it crazed over, so that I could barely see out. I'd read that the emergency thing to do was to punch a hole in it, so that's exactly what I did. And it worked. I now had quite a big hole to see through, so I felt it was safe enough to continue driving home and deal with the situation from there. But then what else could I have done? Mobile phones were still more than a decade away.

The Cortina had a heating system, and a lot of the glass got into it via the slots below the windscreen. More of it disappeared that way when the windscreen was replaced. For months afterwards I had falls of these fragments around my feet at unexpected moments.

12
1980 MORRIS ITAL
A surprisingly good motor

Windscreens seem less of a problem now than they were a few decades ago. I expect that glass technology has improved, along with paintwork and oil seals. Providing and fitting new windscreens was probably a far more profitable business than it is today.

It attracted rogues. I well remember an enterprising scam that was carried out in broad daylight. This was in the early days of mobile phones. The team consisted of three men. Two of them wore clean white overalls, while the third was dressed normally. They would arrive on a big car park with a shopping list. When they saw a model they were looking for drive in they waited until the people in it had walked off to the shops. The man in normal dress then followed the shoppers at a distance, while the two in overalls prepared to get to work. They all had mobiles. When the scout had reported that the car owners were not going to return any time soon, the others set about expertly removing the car's windscreen. The overalls made it all look very professional. No one paid them the slightest notice. They were soon able to put the windscreen in their van and make off.

According to the report in the papers, even on the day when they were caught in the act by a driver who returned to his vehicle sooner than expected, the scout having lost track of him, the team very nearly convinced him that they were carrying out the orders of his insurance company. We replaced the Cortina with a Morris Ital some time in the early 1980s. It's hard to remember exactly when, in fact it's hard to remember much at all about this car, including why we bought it. Both Morris and Austin had by this time become part of the big nationalised company, British Leyland. This was partly to do with what was known as the economics of size. But the new company wasn't the success that had been hoped for. It was prone to strikes, and their cars became a byword for all that had gone wrong with the British car industry. Japanese cars, which were being imported in large numbers, were setting the standards for quality, reliability and everything else.

Can anyone now remember the Morris Ital? It was of the same era as the Austin Princess, although it was more closely related to the Marina, which as I recall came in two shapes: a fastback and a saloon model. A fastback was a car with a sloping rear instead of a step down to the boot. The word was in vogue for a while. The Ital somehow emerged from the Marina saloon. The name was intended to make it sound Italian, I'm sure. Anything Italian would automatically be seen, or at any rate thought of, as stylish.

But having said all that, the Ital wasn't a bad car at all. It was, for example, my first car with a radio fitted as standard equipment. The radio was easy to operate as well. When buying more recent cars I have had to insist that the radio be tuned to my personal requirements before I will sign the papers. There was none of that with the Ital. When I set off from ICI on one of my many long drives on company business, I could tune in to Radio 4 and listen to a play, or a discussion programme, without any fear that thumping, screeching pop music would blare out at me before I could silence it.

I haven't seen an Ital on the road for many years, but I

remember it now as really quite a nice-looking motor, in no way an offence to the eye. The car was silver. I guessed it might soon have rust problems, so the first thing I did with it was to paint a broad band of black underseal around the bottom edges of the bodywork. I had noticed other drivers had done the same. It was perfectly acceptable, in fact rather stylish, and would prove to be effective at keeping the rust at bay.

Yes, I had no trouble with the windscreen on that car. Or with the brakes, or the steering, or the radiator, or the accelerator spring, or even with excessive oil leakage. About the only negative thing I can recall was locking myself out of it one day when I was miles from home.

Not that I should blame the Ital. It was easily done with any car at that time. They didn't have today's fancy keys, with which you can lock or unlock your car door at a distance. The correct thing was to lock them by simply turning the key in the lock. This was the same key used for the ignition. You saw the locking button go down inside the window. I wasn't alone, though, in discovering that you could also lock the car by first pressing down this inside button, then pressing the one on the door handle - which was intended to be used for opening the door - as you closed it. It seemed to save time - five seconds at least. But of course there was a danger. Since you didn't need it for locking purposes, you might easily leave your key in the ignition. People did it all the time.

So there I was somewhere in the Birmingham area, anxious to get off home, but locked out of my car. Fortunately I hadn't left the firm I'd been visiting, and my friends there were ready with advice. Had I left my driver's window slightly open? I told them I had. It was a sunny day, and I often did that so that the car wouldn't have over-heated inside when I returned to it. They didn't yet have air conditioning.

There was a way of opening car doors without a key, one of them explained. You needed a piece of wire, You made a loop on the end of this, bent it in the right place, then threaded it through the window gap. The locking button wasn't actually a button. It had a sort of flanged end. What you had to do was manoeuvre the wire loop over this flange, then pull it upwards very carefully. It needed a little patience, like fishing, but wasn't very difficult.

Well, the man telling me this sounded like an expert, so I asked him to find a piece of suitable wire, and then come out with me and demonstrate the technique. I stood back and watched him fish. It didn't take him long. As we talked he told me that it was a method devised by car thieves.

I was careful not to leave my car windows slightly open after that, even on the hottest days. And I forced myself into the habit of always locking the car with the key.

13
1984 AUSTIN MAESTRO
The voiceless golden chariot

My general experience with the Ital must have been satisfactory, for as the woes of British Leyland worsened I bought another of their cars, later in the '80s. The Maestro was their latest model, possibly their last. In size it was a step up from the Metro, which in turn had been a step up from the Mini. But the Maestro wasn't exactly a big car. Not that I minded. It seemed about right for our current needs, our children now being far beyond the toddler stage. In fact our elder daughter was soon to start at university. Royal Holloway, down in Egham.

A big selling point or gimmick when the Maestro was launched had been that it would be the first car with a voice. It wasn't claimed that you could have a conversation with it, nor was it to be equipped with an early sat-nav system, to tell you how to get to where you were going. It would simply remind you if you had forgotten to fasten your seatbelt, or if a door had not been properly closed. That sort of thing. The voice was female, they said, so that was all right. I didn't fancy a gruff man's voice giving me instructions.

But something must have happened to this voice, for little more was heard about it. Evidently the idea proved to be unpopular, and had been dropped. The Maestro I bought, which was fairly new, had no voice. I confess I was a little disappointed - though not about the verbal reminders I would be missing to put on my seatbelt and the rest. I'd been wondering if the lady might have soothing or uplifting words for you as you set off on a long drive, or when you got back to the car after a tiring meeting. I have pleasant memories of the Maestro, even though I did a stupid thing with it on the day after I bought it. The problem was the radio aerial. This was fitted on the roof, just behind the windscreen. That's common enough today, but the Ital's aerial had been on the wing. Not that I had yet given aerials any thought. The tilt of the Maestro's could be varied, I discovered later, and I had brought the car home with it in the raised position. I wasn't yet in the habit of leaving cars parked on the drive overnight, so I diligently lifted the garage door when it was time to put it away and drove it in. No problems so far. But next morning, when I raised the door and reversed the car out, I heard a horrible, metallic noise. The garage door had a flange on the bottom edge and the aerial tip had got stuck in this. It suffered a severe mangling. Fortunately the garage where I'd bought the car fitted a new one free of charge.

A bad start, but there was no other damage, and as I say I did like that car. Maybe it had to do with the colour, for it had the purest gold finish I'd ever seen. Not a pale sort of gold, or light sort of bronze, but golden gold. It was what had attracted me to it on the garage forecourt. Cars had once all been black, now it was possible to buy a gold one. I won't say it made me feel rich, but it did make me feel good, even if the car didn't have a voice.

We'd had the Maestro for only a year or so when our daughter started university. The drive down to Royal Holloway was much easier than it would have been a decade earlier. Not only had the M1 and M6 been joined up north Birmingham and Coventry, but the M40 and M25 had also opened. I could do almost the entire trip by motorway. In spite of its superior colour the Maestro had to serve as a workhorse, for students need to take a tremendous lot of stuff away with them. It would be piled so high on the back seat that I could hardly see out of the rear window through the mirror. And I seemed to do most of the loading and unloading, while my daughter sought out her friends.

But I enjoyed those long drives with just the two of us. It was an opportunity for her to get her own back on me, however. You could play tapes in the Maestro, and I had been tormenting my family with my vast collection of Frank Sinatra albums, most of which I had taped for playing on long journeys. But now they didn't get much of a look in. My daughter had her own collection of tapes. A band called The Mission was a great favourite of hers at that time, I recall, as also were The Sisters of Mercy. It was all the same to me. I was no more up to date then with rock-and-pop than I am today. Something else I was able to do on my return journeys was to call in to see a writer I knew. This book is by no means my first attempt at writing. I've been at it as a serious hobby for almost as long as I've been driving, though most of my efforts have only been published in obscure literary magazines. I had been in correspondence with the man for some years.

I haven't given the names of the people in this story. That's because it's mainly about cars. The cars are meant to be seen as the characters. If I brought the people too much into the picture, with lots of dialogue, they would very soon take over, with the cars pushed into the background. So I'm going to stick to this policy by not divulging the name of my writer friend, even though in his heyday, fifty years or so ago now, he was quite famous. The titles of two of his novels have become handy phrases for journalists, though both have long been out of print. The earliest of them was filmed by Hitchcock in the '50s. The film was a great success, too.

He now lived close to Milton Keynes, having been appointed as Writer in Residence for the new city back in the '70s for a year. In fact he was still somehow occupying the flat that had gone with the job, the upstairs rooms of an old farmhouse. The downstairs part had been turned into a museum. It was a difficult place to find on my first visit in the Maestro, and not much easier later. I used to get lost in the maze of roundabouts.

His career had declined. I had first written to him in the late '70s about something in a newspaper article of his, and been surprised by his quick reply. He was a complusive letter writer, I soon discovered, who couldn't leave one unanswered. I was the same myself. So we got into a long correspondence. Now, on these return journeys from my daughter's university, a decade later, I had the chance to get to know him a little better. He was living in near penury, I found. I often took him out for a good meal.

I won't give any more clues as to his identity, but leave him as a mystery. You don't expect to have a mystery author in a book about cars. Think of it as a bonus. He was still managing to run a car himself, by the way, an old British Leyland Princess. That's when he could get it started.

14

1990 VOLVO 340

The terrible day when the bottom fell out

My wife had a small company car at her disposal in this period, so I had been tending to make the decisions when we changed our family car. She was a safe and competent driver locally, but she left longer journeys to me, even though her nervousness as a passenger wasn't getting any better. It was in an effort to do something about this that we decided together to change to a Volvo.

Safety was the big selling point with Volvos at the time. They had safety bars in the door panelling to protect the occupants against side impact, for instance. I hadn't heard of that before. I dimly remember a TV advertisement in which a collision was set up. Dummies in the Volvo came out of it unscathed, whereas they would have died in the rival car if they'd been real people. Something like that.

To make my wife feel safer still, we bought the car new. The man in the showroom called after us to enjoy it, as we were about to drive off. I can't say that I ever did. It was a heavy car to drive, probably because of its extra thick bodywork. Yet amazingly, it didn't have power-steering, which was becoming standard by 1990. I had some difficulty in turning sharp corners in it, and had to cut them. I didn't like the colour, either. It was dark blue, but wouldn't shine no matter how much you polished it. I kept these negative feelings to myself, however. As it turned out, we held on to it for longer than any car I've ever owned. But the true reason why I can't look back on it with any pleasure is because it was the car that I drove through the worst period of my life. I don't mean we were in a bad car crash, in which the Volvo's rubust safety features failed to protect us, though you could say that my life - our lives - were in a prolonged crash, metaphorically speaking.

That trouble wouldn't begin till 1998, however. In the meantime it took over as the workhorse for the university trips, a job for which it was well suited. And no sooner had our elder daughter got her Royal Holloway degree than our younger daughter started at Warwick.

But when we got the Volvo in 1990 I was still visiting my friend the writer. He'd had an extraordinary life, and it was time for him to write his autobiography. He was too imaginative for that, however - would I consider doing the job? And so we began a collaboration.

I don't think it's a significant further clue to his identity to say that he spent much of his early life in Cambridge, living in a council house, rather than at the university. His family still lived on the same estate. I remember visiting his sister there one weekend in connection with the book. What I remember best, though, is having a cooked breakfast for three consecutive meals.

When on long journeys alone, unconnected with work, I had started to enjoy stopping at a Little Chef and treating myself to a full English - usually for lunch. There was a concern all those years ago, as there is today, that such breakfasts, though they might set you up for the day, were unhealthy, and could even lead to a heart attack. But we didn't have them at home, and I felt safe enough indulging myself every once in a while.

I used motorways when I could, but I knew by this time where Little Chefs were to be found without having to stray too far from my planned route. And so it was that on my long drive to Cambridge I stopped for one of these generous breakfasts, with extra toast, and of course a pot of tea.

I hadn't bothered to book a room for the night anywhere, as I had made quite an early start and believed I would be in the city soon enough to find a place to stay. But Cambridge is a very long way from Cheshire, and the journey took longer than I expected. I had also been given by phone only fairly basic directions for finding the lady's address. I couldn't imagine that there'd be many council estates in and around Cambridge, but I was wrong. I had to find a shop where I could buy a street plan. The search for the house took me most of the afternoon. I felt pretty sure, though, that the writer's sister would be able to recommend somewhere to stay.

She probably didn't have many visitors, and when I arrived at the house at last she was very pleased to see me. She couldn't think of anywhere at such short notice - would I like to stay with her? She had a spare bedroom. I was tired after the long journey, and said that would be fine - but what about food? There was nowhere very suitable within easy reach, she said, and it was Saturday night, but she could cook a meal for us herself. She hadn't been shopping for a day or two, but would bacon and eggs be all right for me? I said that would be fine, too. I didn't of course mention my earlier meal, out of politeness.

After eating, we talked about her brother's hard childhood for some time, then I yawned and said I would like an early night, if she didn't mind. She'd arranged for me to meet other members of the family next morning, then we could have lunch somewhere. After that I would have to set off for home.

When I came down next morning the lady called to tell me she was busy in the kitchen. I was about to ask for something light, when she put her head round the door and said she hoped it was acceptable, but she was cooking me bacon and eggs again. It would set me up for the day. Our younger daughter started at Warwick in 1993. By then I had finished the biography, which was published without fanfare, and my writer friend had died. Things had also changed at ICI.

When I had joined the development team in the mid-1970s the company was still enormously profitable, so money to fund our various projects wasn't a problem. It was expected that a certain percentage of the profits would be spent on R&D - otherwise the share price might suffer. If there was a problem, it was thinking of ways to spend all this money. Now the situation had completely reversed. Development had gone out of fashion. It was thought to be expensive. Technology was the new in-word. R&D became R&T. And there was a new fashion. We were moving into the age of early retirement, with generous severance terms. Retiring early was made to sound like doing the decent thing. In 1993 I was only fifty-four, but the laboratory where I was based was being run down. I knew I hadn't long to go. For a long time now, using your own car for visiting suppliers and development partners had been forbidden. We had to use hire cars. And these trips were scarcely needed any more. The game was almost up, you might say. But I'd had a good innings. And done useful work, let me add.

I still had the long drives with our younger daughter to look forward to. The Volvo was at least dependable, and they were a relief from the miserable situation at work. Very little in the way of negotiation was needed for me to take the necessary time off. We were in fact almost at the point where there was no one to ask. I had many cooked breakfasts on those trips, not having been put off by my Cambridge experience. She was based in Leamington Spa, commuting daily from there to college. Three years later she also got her degree.

The Volvo wasn't entirely dependable, it turned out, though I had the feeling that it ought to be, since there were so few other things to endear me to it. I didn't often do routine checks. Then one day it let us down. My wife's sister had been taken ill. They lived down in Chelmsford and we set off to visit her one Saturday. We were close to Walsall, on the M6, in slow-moving traffic, when I saw steam rising from the front end. The temperature gauge had moved into the red zone. I pulled onto the hard shoulder.

But unlike in the breakdowns of my early driving days, we were of course in the RAC. We still didn't yet have a mobile, but by good luck we were close to an emergency telephone. It was the only time I've ever used one. We wondered how the rescue vehicle would be able to fight its way through the dense traffic, but we were close to a slip road and didn't have to wait more than twenty minutes. Then we were towed to a local garage, where a new radiator was fitted in a couple of hours. Nor was the charge excessive. We were soon on our way again. It was all so wonderfully efficient. Never since that day have I for one moment begrudged my annual membership fee, no matter how rarely I have to call them out now.

My sister-in-law's illness was serious, but she began to recover. We thought that would be our last health worry for a while. Then the year 1998 arrived.

I had officially retired from ICI, but was continuing to do similar work via an agency. We almost got to the end of the year without health trouble. Then came the day before my sixtieth birthday, just after Christmas. Our daughters would be there with their fiances for a celebration in a local restaurant. But in the evening I began to feel ill. I had no sleep, so many times through the night did I have to go for a pee. Next morning my appetite had deserted me. I made it to the restaurant, but I had to leave early, unable to eat a thing. Severe pains then started. I was bringing up green bile. Soon I was in hospital with a total stomach blockage. Everybody thought it was cancer. My wife was told to expect the worst.

My consultant did an exploratory operation and found what he described as a vascular mass. He wasn't sure what it was, but leaned towards cancer still. I spent a month in hospital, and was discharged still believing I had the disease.

But it wasn't cancer. The abdominal mass disappeared under the influence of routine antibiotics. I began slowly to recover. Then my wife started feeling tired. After a visit to her doctor she was given an appointment to see a specialist. Her blood-cell count was low. I set out for the hospital with her in the Volvo. And the car let me down a second time. The underside cover of the engine compartment fell off - or half fell off. I didn't even know it had such a thing, but sensed that something was dragging along the road beneath us. I stopped to investigate. The cover was made of a cheap material, not unlike painted cardboard. I tugged at it till it came away, then dumped it in the boot. I never replaced it.

It was the start of more bad times, not least for our daughters. My wife was diagnosed with leukaemia. Chemotherapy followed. Things looked bleak, but she got into remission. We had eighteen months of near normal life, but then the leukaemia came back. As also did my stomach trouble. I had another spell in hospital, which included an emergency Saturday night operation for a burst lower intestine. I was in there for five weeks, emerging thinner and weaker than ever - but on the road to recovery again. Meanwhile my wife's condition had worsened. Early in 2002 she went into hospital once more. She never came out. We had great help throughout these bad times from her sister, who had long stays at our house.

Somewhere in that grim period we managed the weddings of our two daughters. We also bought a second car, for my wife's use when she lost her company car. Later, we sold the Volvo. Neither of us returned to work at any point. Work was now history.

15

1996 FIAT PUNTO

The one I was sadly left with

It was in this car, the car intended for my wife's use, that I'd sat outside the Countess of Chester hospital, gathering the strength to go in. Now it was mine. But it was as good as any. The Punto wasn't a woman's car, particularly. Not that I gave a damn about such things. I am not, and have never been, a *Top Gear* type. It was a metallic blue, easy to drive and reliable enough - though I did have to call out the RAC a few times when it failed to start at home in a morning.

But then something started to happen, only a few weeks after my wife's death. My elder daughter had settled close by and I was often a passenger in her Focus. I started to lean towards her as we travelled, my shoulder pressed against hers. She sort of carefully pushed me back, and I'd try to get a grip. What could be difficult about sitting upright in a car? But it would soon happen again. Very odd. Then a neighbour suggested we go for a walk and I kept tripping. I could barely stay on my feet.

I wasn't particularly concerned. I'd been through too much for that. But next, when driving, I thought I detected something wrong with the clutch. It wasn't operating properly, somehow. The steering was also misbehaving. Several times I hit the kerb quite hard. As with the leaning and tripping, I did my best to get a firmer hold of myself, but it would happen again. I blamed the car. Bloody Fiats!

The spring of 2002 had arrived. Our back garden hadn't been

touched for months. I hardly felt up to the task, but thought I would start arranging a few tubs in preparation for planting. It wasn't very difficult, but I fell over, on my back. Our cat was watching, with a puzzled expression. I found I couldn't get up. I was on the ground for twenty minutes, experimenting with various ways of getting to my feet. At last I succeeded, more puzzled than worried. Then I really did fall, banging my head hard on a hard surface.

But I had an invitation to visit local friends that evening, and didn't think of pulling out of the arrangement. On the way I hit the kerb several times. It was a Friday, and my daughter and husband were going away for the weekend, to his relatives in London. I told her I'd be fine, and they should go. I myself would be visiting other friends next day. They lived way up the M6, near Chorley.

I set off feeling not too bad, but I had to pull in at one of the services, suddenly extraordinarily tired. I sat in the car and slept for an hour. Then I carried on. I reached the Chorley house safely and the evening passed without any more trouble. I drove home safely, too, and thought that whatever had been wrong with me had now passed. I got through the Sunday and Monday, concentrating on chores. My daughter was phoning each day after she finished work, and did so on the Monday. We lived in a three-storey town-house, with the living quarters upstairs. She asked what I was doing. I said I was lying on the floor at the top of the stairs, where the telephone was. I didn't know why.

I thought I'd finished with hospitals - but no. First I was sent to Warrington General, then to the Walton Centre, in Liverpool, which specialised in neurological disorders. The betting was that I'd had a stroke this time.

Well, they were wrong again, as with cancer. It was a less serious condition called an epidural haematoma - pressure under the skull. An immediate operation was needed. I emerged from this stay in hospital with half of my head shaved and two dents in the bald part. I wore a cotton hat outdoors all the time, and had to give up driving for a while. Friends very kindly rallied round. It's a curious thing, but I have no clear memory of this period from a domestic point of view. My hair must have gradually grown again, but I don't remember spending much time looking in the bathroom mirror for signs of this. It must have taken several months, and in the meantime the other side of my head would have needed trimming. To go to a barber in such a state would have been embarrassing, if not weird. Did I ever do it? I don't know. Did my daughter cut it, or did I hack at it myself? Again I've no idea. But the summer arrived, and we had a garden party for friends. In the photographs I've got a normal head of hair. Magic.

There was of course nothing wrong with the Punto - either with the clutch or the steering. I had imagined it because of my medical condition. I must have looked a sorry sight walking around our village, in my temporary hat. Or rather stumbling about, for I had developed a painful ingrowing toenail. So I was anxious to get back behind the wheel, where I would be less visible. This had to be negotiated with both my GP and the doctors at the Walton Centre, but eventually I got the all clear. I don't remember any paperwork connected with my period as an enforced pedestrian, but I could have forgotten. I was still wearing my funny cotton hat for the last of these interviews, I do recall. It's a little surprising that there weren't further delays.

I was sixty-three when I lost my wife. It can seem quite young these days, even after you've been ill yourself. I certainly wasn't yet ready for sheltered accommodation, or anything of that sort. And so I did what people do in that situation - I joined several classes aimed at retired people.

That, however, came after I'd tried an organisation designed to help the bereaved. The group met in an old, out-of-use church, in Chester. I arrived late, having had a difficult time parking. There were heavy doors to be pushed open. I could already hear voices inside. The voices didn't sound at all like the voices of the bereaved. I was expecting to find sad people in dark clothing, seated at small tables, sipping a glass of white wine, and murmuring quiet thoughts to one another. But as I approached the voices grew even louder. They could almost be described as raucous. Then I opened the last door - and found lines of people sitting at long trestle-tables. It was like bedlam. They were playing bingo.

I also tried bereavement counselling, but that wasn't really for me, either. So I decided on the classes. I'd been a good artist in my younger days, so I signed up for that. And my wife and I had tried to learn Italian, so that seemed appropriate, too. Later there was also a book club. I dined out a lot with friends as well and I had a go at cooking for guests at home. It was all textbook stuff for a bereaved man, and it continued into the next year. Only I knew it wouldn't do for ever. My wife had been in favour of my meeting somebody else. I might live for another fifteen or twenty years, she'd said, you never could tell. And besides, I needed looking after.

I hadn't so far given that much thought. Nor had I considered changing the Punto. It had of course been bought secondhand. It was now seven years old. A new start perhaps needed a change of car. Another Ford, maybe.

16

2003 FORD FIESTA

The car for a new beginning

Being a widower, living alone except for a cat, isn't really a lot of fun. I had known a few years earlier a man who had lost his wife. He was a small local publisher, who was publishing a book of mine. Visiting one day, I complimented him on how clean and tidy his house was. He told me it was dead. I asked what he meant. He explained that nothing in it ever moved, unless he himself moved it. When he came home it was exactly as he'd left it.

I now felt the same thing in my house. That's why I stayed away from it as much as I could. A house is somehow a woman's territory, the interior anyway. I got some relief from writing stories, which were often published in small literary magazines I had discovered. I used a small bedroom as a study. Our cat used to lie on the bed and watch me at work. Or she'd sleep. She would wake up suddenly, though, if I stopped typing. She seemed to find it reassuring.

My first grandson had arrived, too. But I wasn't yet old enough to see being a grandad as a full life. Had I been a few years older it might have been different. * So I traded in the Punto for a nearly-new Ford Fiesta, at a Ford main dealer. Things had changed since some of my earlier experiences. They couldn't have been more genuinely helpful. Or perhaps I was looking and feeling a little more prosperous. The model I bought has since been superseded, but I prefer its styling. It was navy-blue and polished up nicely.

Of course I didn't set off in earnest pursuit of the opposite sex. I was now sixty-five. But I felt more confident about the new car for long distances. My younger daughter was living in Sutton, south of London, my brother in Farnham, and my sister-in-law in Chelmsford still. So I would put the cat in the cattery, then make a brief touring holiday out of my trips south. It was enough for a while.

The first step I took towards meeting somebody new was to write to a lady whose name I'd picked up in one of the magazines. I made commenting on a story she'd just published my excuse. We got into correspondence. She was a widow. But she admitted to being older than me, and made it clear that she was content with her solitary life. That didn't deter me. She lived in Ludlow, Shropshire, one of my very favourite towns. I suggested I paying her a visit, and taking her out to lunch.

And so we began to see one another. It was a long drive down to Ludlow, but to set off on a journey knowing somebody was waiting to see me - and me alone - beat joining classes any day. I was about to set out one misty morning, I remember, when she phoned to tell me there was a thick fog in Ludlow, and not to come. I fibbed that it was clear in Cheshire, and should soon be gone in Shropshire, according to the forecast. I knew that I wouldn't be faced with smog.

The friendship didn't last. I'm afraid there was more in it for me than there was for the lady. But I was grateful to her for allowing my intrusion into her life for a month or two. She'd been a ballet dancer long ago, and had known Margot Fonteyn. My next move was to put an ad in the *Guardian's* personal columns. Nobody local responded, but then they wouldn't have known where I lived, and I wasn't sure in any case that I would wish to take up with somebody from around the corner - if such a person existed. I liked the idea of operating at a distance, and was almost as interested in where the ladies who answered my ad happened to live as I was in their hobbies and enthusiasms.

Ludlow had rather spoiled me, though. I was rediscovering the old A roads, which by now tended to be empty of traffic compared with the motorways, and the one from Cheshire down into Shropshire had been a delight to drive.

But I met quite a few of these *Guardian* women, sometimes by car, sometimes by taking a train into Manchester. It might have been the way they responded to me personally, but the general impression I formed was that if they were in need of a man in their life it was only part-time. Their lives were already filled with absorbing interests and family commitments. I half expected them to produce a time-table, indicating when I might be fitted in. Men were clearly not their top priority. As for providing TLC, that possibility wasn't on their agenda at all, I could tell. Occasionally, however, I would meet somebody who was obviously interested in me, when I unfortunately didn't feel the same way. Those situations were the hardest to deal with.

I was lucky, though, to be contacted by a lady who lived up in Ulverston, close to the Lake District. That meant a very long drive, much of it on the M6, but that was no deterrent. As with my Ludlow friend, I would set off in high spirits.

We used to meet at Brantwood, John Ruskin's former home on Coniston Water. In order to be in time for coffee, it wasn't feasible to drive all the way up to Ambleside, then make my way back to the lake from there. So I found a new route - new to me - which involved driving up the eastern shore from the south end. Again I didn't mind that. Exploring a part of the Lake District with which I was unfamiliar added to the pleasure of these trips. After we'd had coffee we would spend the day seeing an exhibition, or just driving and walking, with lunch somewhere included. It was apt to be only a short day, as the lady had a granddaughter to meet from school. I had a long drive back, but I didn't care in the least about that. What was a new car for?

We did once plan a trip to London. She would drive down to Cheshire for once, and leave her car on my drive. I would do the rest. That was the plan, but she was very late arriving at my house. Then I got a phone call. She'd made a mistake and had joined the wrong motorway near Preston, taking the M61 towards Manchester. And that wasn't all. Hers was quite an old car, and the exhaust was making a terrible din, she said. She would have to get it fixed, and was now at the Bolton services. Would I come and rescue her?

The quickest route was for me to drive up the M6 to Preston, then down the M61 to the services. That took about an hour. We had a coffee, and considered what to do. I suggested we might take her car to a local garage and leave it with them to have a new exhaust fitted. We could continue with our plan and the car would be ready for her to pick up on our return. It made sense to me, but she wouldn't hear of it. She was afraid the car would be stolen if she left it, and she was dependent on it. She didn't in any case think it was fit to drive any further.

What to do? I was in the RAC. She wasn't. I explained that, as I understood it, the organisation came to the rescue of the driver, not the car. If they towed her car to a garage I could pretend that I was the owner, and we might get it fixed more quickly. To avoid any awkward questions, I moved my car to the far end of the car park.

By now I had a simple mobile phone, so I got onto the RAC. I had every confidence they would soon be there, but they let me down for once. We waited half an hour, then I asked if she would mind starting the engine to see how loud the exhaust noise was. I had certainly heard worse, remembering the business with the old Zephyr. I checked underneath to see that it wasn't going to fall off, then said that if the RAC didn't arrive soon she would have to follow me back to my home. There was a local garage that would fix the problem.

The RAC failed to arrive, so I phoned to cancel the breakdown and we set off slowly for Cheshire. She didn't know how to make her way through the spaghetti junction near Manchester, where the M61, M62 and M60 ring road all converge, so I had to make sure that she was closely on my tail at all times. But we made it. And we got the new exhaust fitted. It took most of the day.

I think we only met one last time after that - up at Brantwood as we usually did. She'd had enough of it. I think I missed the drive, mainly. I mean, what would I be doing otherwise? I'd be sitting in a supermarket coffee shop reading a newspaper, most likely, to get away from my dead home. Older men living alone often do that.

17

2009 FORD FUSION

Comfort, reliability and durability

This history of all the cars I've owned isn't leading where it might have been expected to lead. It's the ambition of the majority of men to try to move upmarket with each change of car, until they have a BMW, Mercedes or Jaguar parked on their drive to impress their neighbours. Yet here I was in this narrative, into my late sixties and not bady off, on my sixteenth car, yet still driving a Ford Fiesta, and a secondhand one at that. How come?

Well, as I see it the cars have got better and better. One of my problems in these later chapters has been that as they have got more comfortable, reliable and durable, it's left me with less to write about than in the days of frequent breakdowns and other problems. I don't need much more in a car than comfort, reliability and durability, and you get all that with pretty basic models these days.

I've noticed something else as well - which reminds of the time I was taken to Blackpool to see the illuminations. This was long before we had a family car, but you could view them from a tram. Most of the ones that had been brought into service for this purpose were just plain, but there was one tram that was all showy, with its own special lights and other fancy decorations. This, naturally, was the one I was determined to ride on, even though there was a long queue for it. So imagine my disappointment when the tram turned out to be just the same as every other tram once you were sitting inside. The external decoration was for the benefit of the people who saw it pass by, not its passengers.

It's a bit like that with cars, in my view. Their showiness is mostly on the outside - which of course is something you see less of than do your neighbours and the people you pass on the streets. Hiring a car once at ICI, back in the '80s, I was expecting to see the usual sort of saloon model waiting, and was surprised to have a sleek sports car at my disposal. I don't want to make too much of this - it was a Triumph TR7, not a Ferrari - but the car did look, well, sporty. But the moment I climbed into it I remembered that Blackpool tram, for the interior was Spartan. Nor was its performance at all exceptional. I thought the Ulverston lady would be my last chance to start a new life with a new partner. I wasn't feeling altogether proud of the way I'd being carrying on. It could easily look like desperation, I realised now, and was that proper for a man of my age? Wasn't it time for the pipe and slippers? There was a second grandchild on the way. Ought I not to be content with the life of a grandfather, and stop all this gallivanting, as my mother would have called it?

That might so easily have happened. But there was one last possibility. I was in touch with another lady, by another route. She wasn't a *Guardian* woman. But I was half inclined to forget about her. She lived miles away again, in the Pennines, almost in Yorkshire. I knew the area, though. My father-in-law had loved it, living only a few miles down the road. I'd often taken him there on our Sunday visits. So I was intrigued. The place was called Saddleworth. You won't find it on the maps. It's a collection of Pennine villages. Well, my daughters suggested I ought to meet this lady, so what could I do? We met in Manchester, in the art gallery coffee shop.

And we were eager to go on meeting. In a very short time I had a new life. We both did. We had been feeling cheated out of our retirement years, so there was lost time to be made up. She was a widow, with no close family ties, or interests that she couldn't drop. That first year, 2006, we had three holidays abroad - in Prague, Italy and the Algarve. It was a great summer at home weatherwise as well. We had breaks in Ludlow, Lincoln and the Lake District.

It wasn't long before we decided to buy a house together. I was lucky to sell my house just before the crash, My partner was less fortunate, so I joined her in Saddleworth, on a temporary basis as we thought. That was now thirteen years ago and I'm still here. Every so often we've got excited about some other town or village where we might settle instead, but so far it hasn't happened. The Fiesta as usual had been clocking up the miles, so in 2009 we traded it in - for yet another Ford. Although the Fusion had been around for six or seven years, it wasn't a well-known model. But it so happened that we often passed one, parked on the main road in the next village. In truth I didn't at first know the name of the car. Then we began to spot them elsewhere. Usually there would be an elderly man at the wheel with his wife beside him, tootling along. It became a joke with us. We called it the perfect car for wrinklies. It resembled the Fiesta, but was less sporty looking.

Then we visited our nearest Ford main dealer and there was a new, silver Fusion in the showroom. But there were also a Focus and a Mondeo to consider. It happened, though, that we had a very short drive, back home. Our Fiesta actually stuck out into the pavement area when I parked it. We couldn't accommodate a longer car. So we had a closer look at the Fusion. It was the same length as the Fiesta, and we found we liked it in other ways. It was a taller car - that was the main difference. You didn't have to lower yourself into it. We were beginning to appreciate that sort of thing. The seats were very comfortable, too. And the radio and CD player were simple enough to operate. In short, it was the ideal car for us. So without further ado we bought it. Let them dismiss us as wrinklies if they liked. We knew better. With a new car it was time for more travel, and for the discovery of new places neither of us had known in our previous lives and could therefore think of as ours. An early discovery was the amazingly well preserved town of Tewkesbury, close to the Welsh border, between Worcester and Gloucester. It was much in the news in the year of the great floods in that region, its abbey left standing on an island. Yet we visited it soon afterwards and it looked the same as ever to us. Apart from its own charms, Tewkesbury is well placed for visiting the Cotswolds.

Another town we discovered and grew to love is Lichfield, the birthplace of Doctor Johnson. This was by accident. There was no handy motorway south from Saddleworth, but we discovered an A road running down through the beautiful scenery of the Derbyshire and Staffordshire dales. By this route, we found we could join the motorway system south of Birmingham. But the traffic was normally light, so it was well worth using for its own sake. It begins in Buxton and passes through Ashbourne and several quaint villages, before reaching the cathedral city. We stopped in Lichfield that first time for only a light lunch, but we've since had longer breaks there.

Buxton is in the Peak District, and there was that whole area, fairly close to home, to explore. I didn't know it well, even though my father's roots were there, in the Hope Valley. On the wall in a pub we found framed old census records going back to

Victorian times, with many Darwents listed. Our favourite village in the Peak District is Ashford in the Water, between Buxton and Bakewell. It's another place we keep returning to.

To the north we found Clitheroe, a hilly little town where an attraction for us has always been its secondhand bookshop. It's close to Bowland, which was itself a discovery. People flock to the Yorkshire Dales and the Lake District, and miss this beautiful area - sometimes known as the Trough of Bowland, or the Forest of Bowland. Only a mile or two north of Clitheroe is the exquisite village of Waddington, with a stream and lovely gardens running alongside its main street. There are seats, too, so you can enjoy them at your ease. But the thing about Bowland is its quietness. You can drive slowly around the lanes and villages without fear of a car behind you sounding its horn. There is talk of fracking being started soon in the area, but I feel confident the glorious Bowland countryside has its many determined defenders.

The Fusion, meanwhile, was standing up well to all this travelling. It rarely gave us any trouble. But I do remember one extremely cold morning - and the temperature in these parts sometimes descends to minus something in the teens in mid-winter - when I lowered the window to collect a ticket for a car park and it refused to rise up again. We had to drive for a couple of days with an open window until I could get it fixed. It reminded me of my early driving days in the '50s, when driving like that was normal. I didn't feel it necessary to make hand signals, however.

Then, one New Year's Eve, we drove up to a pub on a hilltop for a meal in heavy snow, and I reversed into a dry stone wall in the dark while parking. In our part of the world there are dry stone walls everywhere. They're difficult to miss. But luckily the dent wasn't very big. It got on my nerves, though, then one day, when we were off on a trip up to Bowland, I hesitated before venturing onto a large roundabout with fast moving traffic, and got rammed from behind. The damage this time didn't exactly coincide with the earlier dent, but when I got the car back from the garage the dent had gone. I was assured by my insurance company that with such collisions it was always the fault of the driver behind. This view was contested, however, and it took about a year to sort it out. I still get messages from opportunistic solicitors offering further help with the case.

The funniest thing that happened with this Fusion - though it seemed anything but funny at the time - involved crows. They're intelligent creatures, as anybody familiar with Aesop's Fables will know. They also have a weapon - namely their droppings - and they don't see human beings as friends. There's a bus stop down the road from here, beneath tall trees in which the local crows like to make their nests. The pavement round about this bus stop can be white when they're nesting.

Well, we were on holiday at Nidd Hall in Yorkshire one spring, not so long ago. The car park could get very full. On the evening before we were to leave I wanted to park close to the entrance for convenience, with our luggage to carry to the car next day. There was one single inviting space, under tall trees. But I thought no more about it - until morning. The car was absolutely splattered. I had an old towel in the boot, but could do little but spread the stuff around with it, although I managed to get most of it off the windscreen. We could only think of stopping at a supermarket we'd seen and buying several packets of wet-wipes - the large type, intended for cleaning floors. The job must have taken us an hour, at least. People passing by wondered what on earth we were doing.

18
2012 FORD FUSION
Past, present and future motoring

Crows are intelligent creatures, but I don't believe the bombardment of our car at Nidd Hall was in any way a comment on the Fusion itself. I'm sure they would be just as happy to bomb BMWs or Jaguars, and perhaps even to prefer it. Birds of all kinds seem to like using parked cars as places to deposit their deposits. They're particularly attracted to vertical surfaces, like side windows, you may have noticed. Quite how they manage that, I don't know. They do it in secret.

The car was almost three years old and I wanted to change it, rather than get into requiring MOTs. We were considering buying a Kia Soul next. As the years pass, it must be ever more difficult for car designers to give their new models a distinctive look. It strikes me as not unlike the problem faced by the designers of women's clothing. Women have two arms, two legs, etc. How on earth can new ways be dreamt up to cover them, year after year? In a similar way, cars have a radiator grille, lamp clusters, doors, windows, wheels, and so on. Yet the designers of the Soul have managed to come up with a distinctive look, just like the fashion designers.

It happens, however, that Kia share premises with Ford at the main dealer where we'd bought the Fusion, and they didn't have a Soul in the showroom. I'd had reservations about the name Soul, in truth. I don't like soul music. It gets linked with jazz these days, and it isn't jazz. So we had an opportunity to forget about the Kia Soul. And by the series of logical steps which we had followed with the previous purchase, we decided once again that the best new car for us would be… a Ford Fusion. They were being phased out, but it so happened that one would soon be coming in. They were still popular with older people in particular. Did we want it? It might be our last chance. We said yes - what colour was it? Silver, the man said. The same as the Fusion we were trading in. As the year 2021 approaches, this is the last chapter of the story of my cars. I've been driving for nearly sixty-five years so far… It's time to look back to see if there's anything I can add - any more amusing anecdotes, for instance. Amusing in retrospect, that is.

Yes, I remember a night back in the late 1950s - a Christmas Eve, actually - when the old Austin Cambridge let me down. Not that I was alone. I'd driven with two mates to a dance at Runcorn. You could get there either by the transporter bridge, or via a country road on the Cheshire side of the river. We preferred the latter route, and there were no problems till we were on our way home, very late, when the car broke down. We couldn't attempt to do much about it in the dark, but by a miracle a car came along the quiet road and we successfully thumbed a lift. I remember it was one of the earliest Ford Consuls, quite small, but the three of us squeezed in the back.

So what to do about the Cambridge? The father of one of the lads had a car of similar age, and we persuaded the man to help us tow the abandoned vehicle back home. He was an outwardly quiet man, but with a furious temper. Reversing his car out onto the road, he demolished a fence, but he didn't seem to notice. We travelled in absolute silence. He'd brought a rope, which he secured sensibly to his own vehicle, then tied it round the bumper of the Cambridge. I wasn't about to protest.

If you have ever been towed with a rope you will know what happens. The rope doesn't remain taught. Sometimes you tend to catch up with the vehicle in front - which then begins to draw away as you slow down, and you feel a powerful tug as the rope tightens once more. That happened many times on our homeward journey, with the result that the Cambridge's bumper had been tugged all out of shape. But the man untied his rope, threw it into his car and went home.

The bumper was beyond repair, but my father said he wasn't going to pay for a new one. That was up to me. The man who had towed us had claimed that his car had been damaged, too, and my father had had to pay him something in compensation.

I didn't have the money to buy a proper replacement, but a man I knew in the sheet-metal department at work offered to help.

He made a new bumper for me. It was rather crudely shaped, but with a couple of coats of aluminium paint it looked all right. I must have driven the car for a year or more after that, and nobody ever commented on its funny bumper. I don't believe it was ever even noticed. Life was once like that. Reversing very carefully out of parking spaces is something I seem to have to do a great deal of these days. Talk about needing eyes in your backside! I don't know how anyone with a stiff neck could do it. Supermarket car parks and similar never existed in my early driving days. You can be slowly pulling out of a space, and other drivers will whizz by behind you. And pedestrians, pushing their trolleys, do the same. They always watch the car, not the driver - as if it has a life of its own. If the car pauses, they think it's safe to walk by - never mind that the driver is watching his front end, so as not to scrape the car in the next space. You just know that one day there is going to be a collision. And yet, though I've carried out this manoeuvre hundreds of times, it's never happened at a supermarket. But I do remember a reversing collision on a near-empty car park long ago. I think it's that experience that has left me so very nervous when doing it. It came as such a shock.

It was in my early ICI days at Runcorn, when I was driving the Hillman Husky. I had been working overtime, and when I emerged from the offices the large car park was empty. The only obstruction was a single cast-iron lamp-post, right in the middle. It was daylight still, so this was easy to see, despite being painted in a dull colour. The car park had what I'm sure were intended to be an entrance and an exit, though there were no signs, and drivers took their pick, especially when there were no other cars about.

Well, I made my choice and reversed at some speed from a space at the edge of the car park, turning the wheel so as to point in the right direction. Then for some unknown reason I changed my mind. I would use the other exit. So I swung the wheel again and reversed some more. Suddenly there was this terrific bang behind my. Yes, I had hit the damned lamp-post. I could probably have driven around blindfold for half an hour and never touched it. The car was luckily still driveable and the radiator at the rear apparently undamaged. That would have been when the radiator problems with the Husky first started, however.

The moral is: never reverse at speed. I never do. Usually I am barely moving at all. Among the many changes I've seen on the roads in all this time, a few stand out. Drivers can be more courteous these days. If you do them a little favour, say by letting them in from a side road, they'll give you a little wave behind. And yet they have this constant urge to pass you at the slightest opportunity. I've noticed this particularly in the south, when I've been visiting my daughter and young family down there. They see roundabouts as an overtaking opportunity. You can be undecided as to which of two lanes to take as you approach one, knowing that the driver behind will be watching you like a hawk. You may delay your decision till the last moment, but you can bet that he or she will be on your inside or outside in a flash, then risk everything to be first on the roundabout. Yes, women drivers do this, too, I'm sorry to say. I half suspect they believe male drivers should show them the same courtesy on the road as they do by stepping aside to let them pass on the pavement.

The main thing I notice on the roads and car parks now is the excellent condition of the cars themselves. We have been passing through a painful recession and the nearby towns in this region have suffered more than many. Yet almost all the cars look strangely new, even when you know from their number plates that they're at least ten or twelve years old. Where is the rust? Where is the evidence of the wear and tear from having been driven thousands of miles? You never see the modern equivalent of the old Austin Cambridge. And you certainly never ever see a car with a home-made bumper. As for the future of cars, unless there is finally a rebellion against this relentless digital takeover of every imaginable human activity, there soon won't be such a thing as driving as we still know it. The whole thing will have been taken out of our hands - all the way down to our choice of music as we're carried along, I shouldn't wonder. It doesn't sound like a lot of fun. But perhaps I'm being too pessimistic. I take comfort from the fact that automatic transmission has never really caught on in this country. We like to be in control of our cars, even preferring to change gear manually. So I hope the driverless car, and all that goes with it, is laughed off the road.

Our second Ford Fusion is now eight years old, but I mean to keep it for as long as I continue driving. It has everything I need. I'll soon be eighty-two, so it can't be for very long. Even so, the future of motoring may possibly have arrived fully by then. The next few years could be very interesting. At least I'll be able to write a sequel to this book, telling how, as an unreconstructed technophobe, I've been coping behind the wheel.

About Brian Darwent

The author is a former draughtsman and development engineer, who has also been writing stories and articles for literary magazines for much of his adult life. He has published several books.

Available worldwide from Amazon

---

www.mtp.agency

www.facebook.com/mtp.agency

@mtp_agency

Michael Terence
Publishing

Printed in Great Britain
by Amazon

70395091R00061